Synthesis Lectures on Engineering, Science, and Technology

The focus of this series is general topics, and applications about, and for, engineers and scientists on a wide array of applications, methods and advances. Most titles cover subjects such as professional development, education, and study skills, as well as basic introductory undergraduate material and other topics appropriate for a broader and less technical audience.

Deandra Cassone

Leadership Concepts for the Engineering Mindset

Springer

Deandra Cassone
Kansas State University
Manhattan, KS, USA

ISSN 2690-0300 ISSN 2690-0327 (electronic)
Synthesis Lectures on Engineering, Science, and Technology
ISBN 978-3-031-67797-7 ISBN 978-3-031-67798-4 (eBook)
https://doi.org/10.1007/978-3-031-67798-4

This Springer imprint is published by the registered company Springer Nature Switzerland AG
The registered company address is: Gewerbestrasse 11, 6330 Cham, Switzerland

This book is dedicated to my three children, Kasey, Kristina and Victor, and their families. Having a healthy, happy family is one of life's greatest achievements.

Preface

Engineers have a bright future, and potentially brighter than they may realize. This book begins by providing background on core engineering knowledge and skills and then discusses how to translate those skills "from the factory floor to the board room." The book provides some quantifiable information with regards to core engineering skills and knowledge and how it can translate to the skills and knowledge of senior leaders in organizations. Gaining a broader view of yourself as an engineer and your work environment provides a broader environmental context to apply engineering skills and organizational vision. The book discusses skills and knowledge inherent in most engineers and identifies areas for further development and training to gain a broader understanding of your organization and skill development that will help you reach your career goals, whatever path you choose to take.

Writing this book is driven by the fact that I love seeing engineers leveraging their core skills, knowledge and work styles to succeed in an organization. My career has spanned the corporate environment, consulting and academia, and I have seen engineers do things right, and then also, not-so-right. I believe that if engineers broaden their knowledge in some key focus areas and develop their skills from a broader, organizational perspective, they can open opportunities for expanded leadership positions. The topics in this book are topics that I teach in my graduate engineering management, operations research and systems engineering graduate engineering courses, but are those areas that I've seen first-hand, based on my career experiences. Personally, I see these concepts as critical concepts that engineers should know to better further their careers, no matter whether they pursue a technical or managerial career path.

This book only touches the surface of the topics covered in the various chapters. Engineers can identify resources to learn more and better understand these topics in depth. This can be accomplished by additional graduate courses and degrees, short courses, certifications, reading books and textbooks and mentoring relationships to gain this knowledge. I don't believe that you will waste your time by learning more about these subject areas.

From my upbringing and background, I have been surrounded by engineering leaders my entire life. My father was an academic, business and community engineering leader. Many of his students have become CEOs of companies, senior vice presidents, presidents of universities and successful business owners. He ran two consulting companies, commercial real estate and real estate development companies and was heavily engaged with the community and family. Having functioned as an engineering leader in industry and academia, I personally know those skills and knowledge that I've either learned the easy way, or the hard way through mistakes. I know that engineers can be very successful, even if they may not currently possess and need to develop additional skills to further their careers. It can be done.

Interesting, also, is that many of these executive engineers may not be as widely known as some of the technology leaders that may receive more of the spotlight. Individuals on your alma mater's engineering advisory council or in your university's engineering hall of fame are some of the bread-and-butter types of engineers that have been successful in leadership positions. Beyond that, many engineers that are successful engineering leaders don't receive major recognition even though they are equally successful engineering leaders. They just may not care about the recognition and would much rather focus on the work.

The goal of this book is to point you in the right direction. I have personally seen how each of these topic areas are useful in my career and the career of other engineers and helped me meet my personal career goals. If the book can enlighten you in a few areas and, equally as important, help you to avoid some mistakes, then I would consider that the book has served its purpose.

Manhattan, KS, USA Deandra Cassone

Acknowledgments

I would like to thank my son, Victor Cassone, for reviewing the first draft of the book and providing his input. He is a young engineering leader in the making, and his thoughts and insights were helpful in shaping the way concepts were presented in the book.

I'd like to thank my father, Frank Tillman, Ph.D., PE, for demonstrating engineering leadership qualities, being a role model and exposing me to highly successful engineers at an early age. I'd also like to thank him for demonstrating a strong work-life balance and reinforcing the importance of family along with achieving career goals.

Contents

Exploring Leadership from an Engineer's Perspective

Abstract

Engineers possess skills such as judgement and decision making, complex problem solving and critical thinking that can serve them well in all levels of management in an organization. Retroactive data is usually used to show the number of engineers in management and leadership positions, such as Business Insider's statement that "33% of the S&P 500 CEOs' undergraduate degrees are in engineering" (Aquino in Surprising facts about CEOs, 2011 [1]). This chapter explores data from the US Bureau of Labor Statistics, specifically, O*NET data, that shows the skills, knowledge and work styles of engineers inherent in their educational training and technical engineer positions. These characteristics can be leveraged and developed for higher-level engineering, general management and executive organizational positions, should engineers choose to pursue those positions. Additionally, this chapter discusses technical and managerial career paths for engineers. This chapter forms a foundation for identifying skills, knowledge and work styles for engineers, engineering managers, general managers and executive level positions and identifying similarities and differences between these job classifications.

Engineering Leadership at All Levels of the Organization

Introduction

Engineering skills can be leveraged in leadership positions at all levels of the organization. Engineers may stay in their technical disciplines, become technical leaders or pursue technical or non-technical leadership positions in an organization. This chapter contains data that can be assessed to identify key skills, knowledge and work styles that engineers

D. Cassone, *Leadership Concepts for the Engineering Mindset*, Synthesis Lectures on Engineering, Science, and Technology, https://doi.org/10.1007/978-3-031-67798-4_1

typically have within their areas of technical discipline and in management and leadership positions in an organization. Engineers can then evaluate those characteristics that they do have and identify areas that they lack.

This book is written based on years of corporate, consulting and academic work experience and years of teaching engineering graduate students. Topics addressed in this book are key areas that provide value to engineers in their career development. The objective of the book is to identify specific areas of focus in engineers' career development and provide information that engineers may use from a quantifiable and executable perspective.

Engineers have a mixed view of the term leadership [2]. They may think that leadership is an indiscrete term that applies to those in positions with a less technical background. They may also think that to be in leadership positions, you must lose your engineering objectivity to succeed. Additionally, they may think that moving into a leadership position immerses you in corporate politics. Parts of all the statements above may be true, however, even in technical positions, engineers become technical leaders that provide valuable contributions to an organization. Additionally, engineers' training background translates into skills that are needed at the highest levels of an organization.

This book discusses the skills, knowledge, and workstyles that any engineer should know. This book is not intended to address the breadth of skills that are needed to become a leader at the highest levels of an organization, but to focus on skills that the senior engineer, engineering manager and mid-level corporate executives need to understand to make their career path more meaningful to them. Many great engineer-to-CEO books, lectures and examples exist and provide great information and knowledge for engineers to study. This book focuses on some of the more tactical transition items that engineers should know to better understand how they can excel in their current positions.

Data, research and observations are not always readily available with concrete conclusions to address leadership positions for the mid-level engineer. A variety of sources have been assessed and aggregated to attempt to quantify the anecdotal knowledge on engineering leadership development that seems to be known but not well documented. It has been interesting to find data, from sources such as O*NET, that appears to make the anecdotal more concrete. Understanding the entry level, mid-level technical, managerial and executive position qualities, skills and knowledge is not an exact science. The intent, however, is to provide information when available for you, the engineer, to decide on its value. As engineers, you will inherently evaluate the information presented in this book and decide whether it is useful to you. Look, however, and see if the information presented is something that you already know or is information that will provide value to your career development. Additionally, the book attempts to address why this information is useful to you.

Fundamentally, engineers need to understand a broader view of their positions in a company. Engineers typically feel great reward when they successfully deploy their knowledge to deliver designs, models, projects, process improvement or successfully manage areas related to their training. The remainder of the organizational operations may or may not be of interest to you. A better understanding of the context of your responsibilities within the overall context of an organization will only make you better at what you do. This book addresses some of the key concepts and organizational aspects that an engineer should know to be more effective in their organization.

You're in the Right Field for Growth

From a big picture perspective, it is important for you to have a basic understanding of your industry and how it fits in the projected growth spectrum of all industries. The North American Industry Classification System (NAICS) creates and maintains a common industry classification system [3]. NAICS industry classifications have similar production processes classified in the same industry. NAICS publishes information on inputs and outputs for the various industries so that this data can be used to measure the economy consistently. These classifications are grouped into sectors such as Utilities, Construction, Manufacturing, Retail Trade, Finance and Insurance, Professional, Scientific and Technical Serves, Healthcare and Social Assistance, and several other classifications.

Figure 1 shows employment growth projections for the various sectors from 2022 to 2032. This information is helpful to see key areas where projected employment appears to be strong. Financial analysts and economists at a minimum use this type of data to develop financial and economic projections. It is helpful for engineers to know their industry sectors and to understand whether their industry is positioned for future growth. Much more data is available regarding these types of analyses and can be studied in detail to determine if your current industry sector can provide longevity in the industry choices for your career.

As you can see in Fig. 1, healthcare and social assistance tops the list, but also, sectors such as management of companies and enterprises, professional, scientific, and technical services, transportation and warehousing and information are all very close in terms of the projected change in salary and employment through 2023. All these fields are those that engineers may function in and are top fields anticipated to grow.

The Bureau of Labor Statistics (BLS) develops employment estimates by major industry on a yearly basis. Their estimates show that in 2021, nearly 15% of all employment was in the industry classified as Professional and Business Services. Focusing on professional and business services, this industry had the most employment of any major industry in 2021. Employment in professional, scientific, and technical services sector are projected to grow the fastest. Many engineering positions function in this sector, however, engineers are employed in nearly all the NAICS sectors. The professional and business services sector consists of three industries: Professional, scientific, and technical services, Management of companies and enterprises, Administrative and support and waste management and remediation services. The professional, scientific, and technical

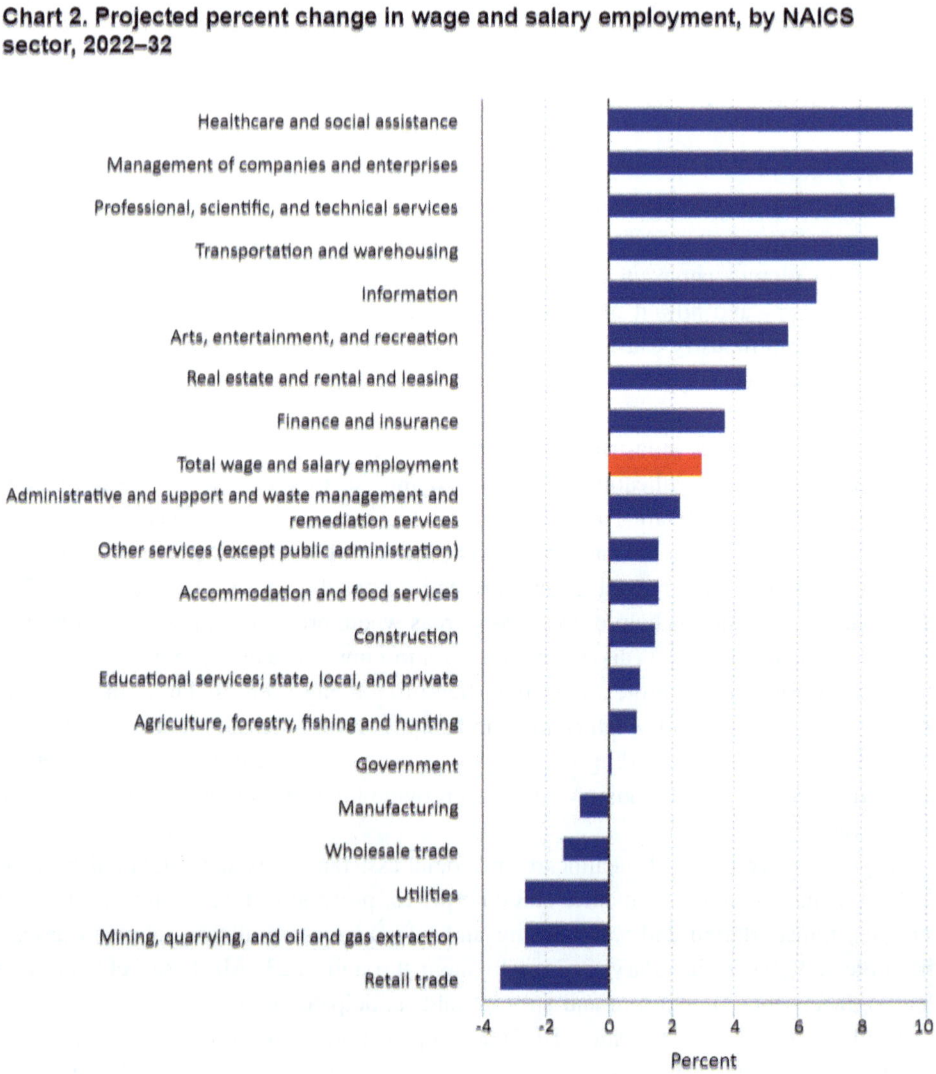

Fig. 1 Bureau of Labor Statistics—NAICS projected change in salary and employment, 2022–2032 [4]

services industry is also projected to grow much faster than the average for the total economy with about seven out of every ten new jobs expected to be created in professional and business services over 2021–31. Figure 2 shows projections through 2031.

Breaking down further the professional, scientific, and technical services sector, four industries are projected to have employment growth greater than 12% from 2021 to

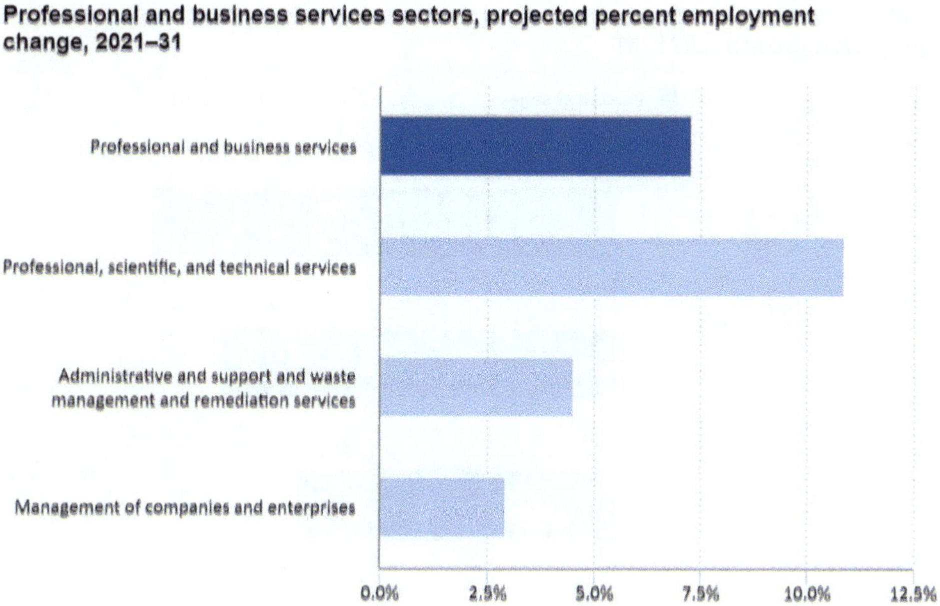

Fig. 2 Professional and business services sector projected percent employment change, 2021–2031 [5]

2031. These industries include computer systems design and related services, veterinary services, management, scientific, and technical consulting services, and research and development in the physical, engineering and life sciences industries. Again, engineering skills are well suited for these high growth areas (see Fig. 3).

Fundamentally, this information is presented to show that engineering and technical skill-based jobs are projected to continue to grow in the future. Engineers and other technical professions have many opportunities for long and meaningful careers. Engineers have a bright future, so thinking through your industry, sector and its growth is helpful in positioning yourself in the right industry sector.

Fundamentally, engineering and technical skill-based jobs are projected to continue to grow. Engineers have many opportunities for long and meaningful careers.

Job Occupations Compared

Finding quantifiable information about engineers and potential engineering career trajectories can be difficult [2]. From an engineer's perspective, it is helpful to have specific information that provides insights into areas needed for career growth. An interesting source of information is that provided by the O*NET Online system provided by the US

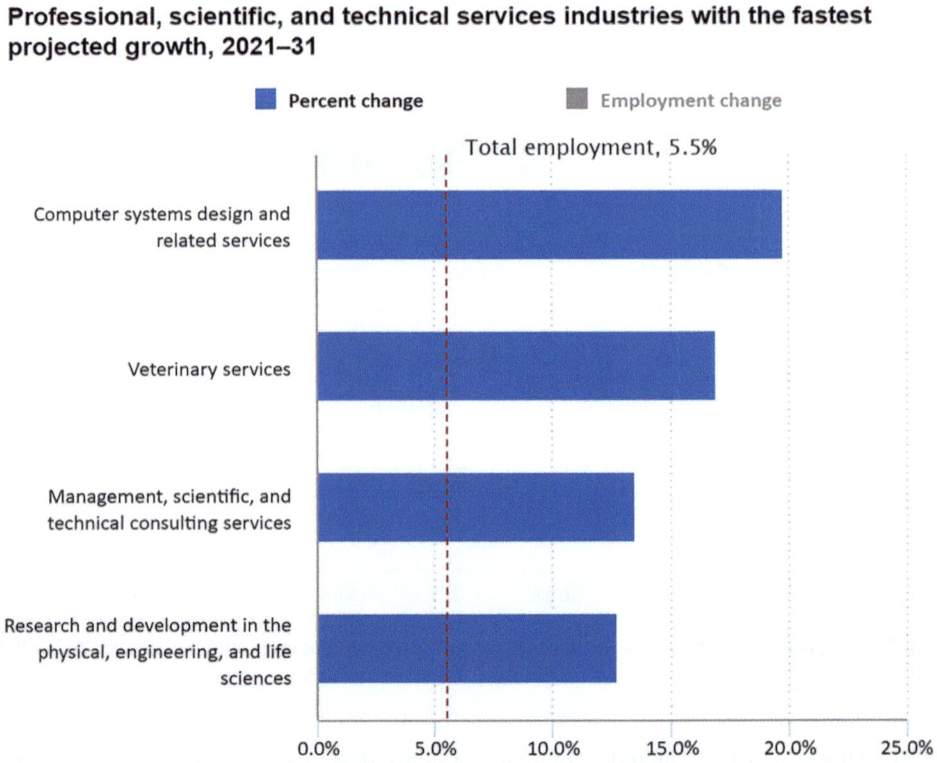

Fig. 3 Professional, scientific, and technical services industries with the fasted projected growth, 2021–2031 [5]

Department of Labor. The O*NET program is the nation's primary source of occupational information. The O*NET program provides valid information about the U.S. workforce and an understanding how to build and maintain a skilled workforce. The O*NET database contains standardized and occupational descriptors on almost 1,000 occupations covering the U.S. economy. This information is continually updated by a broad range of workers in each occupation from job incumbents or occupation experts.

The O*NET Content Model describes a mix of knowledge, skills, and abilities, performed by job occupations. The Content Model provides a framework that identifies the most important types of information about work. These descriptors are organized into six major domains, which include worker characteristics, worker requirements, experience requirements, occupational requirements, workforce characteristics and occupation-specific information. O*NET provides comprehensive information that can be used to understand the work characteristics of engineers, engineering managers, general operations managers and executives. Figure 4 shows the elements of the O*NET Content Model.

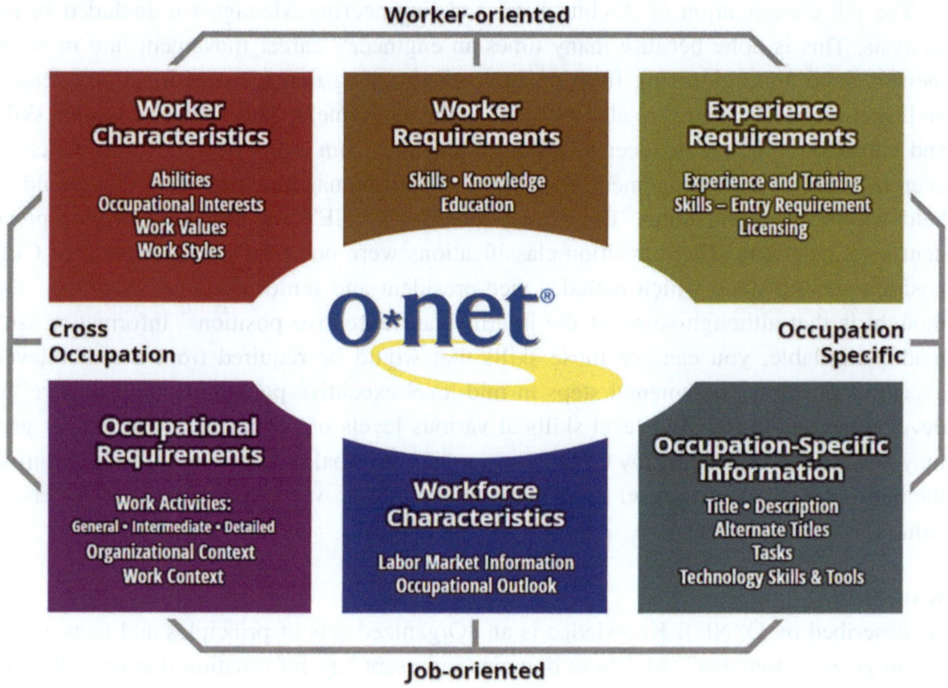

Fig. 4 O*NET Content Model [6]

O*NET data was chosen as a base point for identifying the skills and knowledge engineers have and will need as a basis of career development. The intent here is not to mire you with details about specific data sources, but to identify skills, knowledge and work styles that you can evaluate to understand some quantifiable base skills that you have as an engineer and to identify areas that you may lack so that you can pursue further career development.

Five engineering disciplines have been included in looking at these characteristics in detail which include Mechanical Engineering, Electrical Engineering, Civil Engineering, Chemical Engineering, and Industrial Engineering. These engineering disciplines were chosen to represent a breath of engineering functions which can focus the discussion on characteristics across engineering disciplines. The goal is for you to see the trajectory of skills that are held by staff engineering functions and those that evolve into management and upper management positions. As you will see in the analysis, the primary knowledge, skills and work styles of the engineering disciplines have many similarities and differ in their focus areas associated with the specific disciplines. Detailed information has been provided in Appendix. This section summarizes the findings from the more detailed analysis.

The job classification of Architectural and Engineering Managers is included in the analysis. This is done because many times an engineer's career movement into management is from an engineering function to an engineering management function. General and Operations Managers are also included in the assessment. This shows additional skills and knowledge that an engineer would need to move from a purely engineering function, even in engineering management, to a more general management position that would be held within an organization. The author searched O*NET for director and vice president level positions. These position classifications were not available; however, the CEO position was included which includes vice president and senior executive positions. The thought is that although some of the intermediate executive positions' information isn't readily available, you can see those skills that would be required from a senior level-executive position. Incremental steps in mid-level executive positions would involve the development of those CEO level skills at various levels of proficiency. Your career goal may never be to be a company CEO, however, if your goal is to be a mid-level executive, the more of the executive level skills that you can master will help you provide leadership value to your organization.

Knowledge

As described by O*NET, Knowledge is an "Organized sets of principles and facts applying in general domains" [6]. These domains represent key information that an individual in the position should have.

Figure 5 shows a comparison of the base engineering disciplines and knowledge areas required by management and executive positions. (See Appendix "Definition of O*NET Knowledge Areas, Skill Areas and Work Styles" for a full definition of the knowledge areas.) Areas highlighted in orange represent engineering knowledge areas. Areas highlighted in green represent management skills that aren't typically possessed by engineers from a functional engineering perspective. This provides some quantitative data to show engineers areas that they may need additional knowledge, skills and training.

Looking at the Knowledge category and comparing between engineering, management and executive functions provides an interesting view of the capabilities that engineers possess at a fundamental level and how that knowledge translates to management and executive positions. This also helps highlight areas where engineers need to gain knowledge to successfully move into management positions. O*NET has more than ten different knowledge areas for most positions, but only the knowledge areas with the highest percentage of inclusion by respondents have been included.

Understandably, key knowledge areas for engineers include Design, Engineering Technology, English Language, Computers and Electronics, Customer and Personal Service, Mathematics, Mechanical, Physics and Production and Processing. To a lesser degree, we see Public Safety and Security, Building and Construction, Education and Training and Law and Government. These skills represent core engineering functions that are

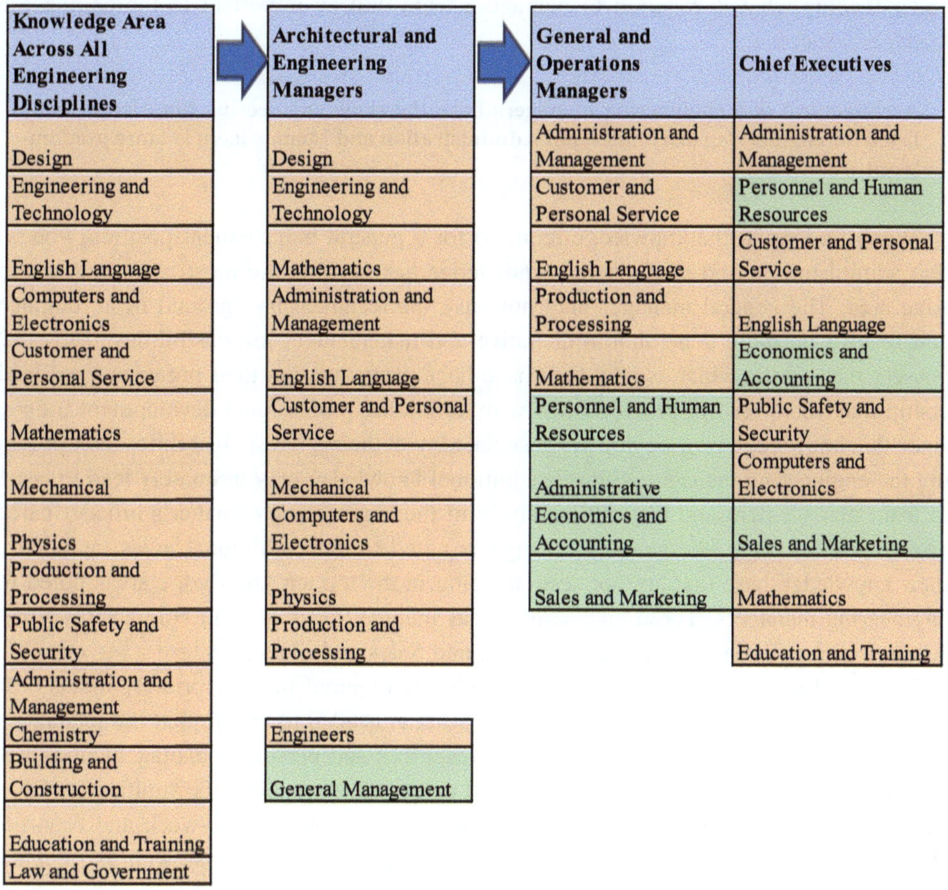

Fig. 5 Comparison of knowledge areas between disciplines [6]

included in their academic and typical engineering specific positions. These knowledge areas represent core skills that engineers will use in their career positions.

As an engineer moves into an engineering management position, Administration and Management is included in the list at a higher level of importance. This supports the idea that many times, as engineers are promoted, they add administration and management to their knowledge base, but in many ways, operate in a similar capacity to their engineering employees with the additional responsibility of managing their team. Knowledge about their responsibilities and requirements associated with the additional team responsibilities may or may not be gained, because the assumption is that they are simply doing similar work but now managing individuals in addition to their previous functions. The lack of this additional training can cause issues in their team management functions. Truly understanding what is required to manage a team will be important to the engineering manager's success.

Architectural and engineering managers have the same engineering knowledge areas listed for engineering disciplines, but Administration and Management is more predominant.

When looking at the knowledge required for a general management position, you see that administration and management knowledge has become the most important knowledge area. The general manager may not have the technical background of an engineer and focuses on more general administrative and management responsibilities. They may possess these skills which is why they have been promoted, but their organizational leadership may more readily provide them with additional professional development training since this knowledge is more directly associated with their job responsibilities. Not providing the engineering manager with this additional knowledge may ultimately lead to issues in team management and may ultimately limit the engineering manager's broader career growth. Additionally, the general manager is noted to have additional areas included in their knowledge base that are not seen as being in the top ten knowledge areas noted for engineering managers. These knowledge areas include Personnel and Human Resources, Administrative, Economics and Accounting and Sales and Marketing.

Personnel and Human Resources knowledge is required to appropriately manage the engineering manager's team and understand what is needed to ensure that the manager is following company policies around human resources and personnel management as well as understanding how to provide growth and career development opportunities for his or her team. The manager should also know the less glamorous paperwork and reporting functions required for their department. Economics and accounting as well as sales and marketing can be far more foreign to an engineer. Business degrees provide much more focus on these areas as they develop managers. Most engineering course curriculums focus much less on these areas or may not address them at all in their course curriculum. Engineers may need to seek out gaining this knowledge through advanced business degrees or professional development opportunities. These are knowledge areas that are important to success in career development and should be learned. Later chapters of this book will address some important high-level concepts in these areas; however, entire fields of study exist and can provide details and insights into these areas.

Knowledge areas for General Managers include Personnel and Human Resources, Administrative, Economics and Accounting and Sales and Marketing. These knowledge areas are not predominant for engineering managers.

From a career trajectory perspective looking at Chief Executive skills, a higher level of importance is in areas of Personnel and Human Resources, Economics and Accounting and Sales and Marketing. You do see that knowledge areas that are evident in the engineering disciplines are listed for the chief executive positions but are not as pronounced as that of an engineer or engineering manager. Knowledge is key and to be successful in executive level positions, the engineer needs to gain this additional knowledge to equip him or her in these higher-level positions.

The knowledge areas for the chief executives are much broader and more comprehensive from an overall corporate perspective rather than from an engineering staff or management perspective. You'll see a theme throughout this book with regards to leveraging basic engineering knowledge and skills at a higher level of responsibility and sphere of control. To move into management and leadership positions an engineer must broaden the domain with which their core skills are applied. The author believes that the engineering knowledge and skills applied to a broader domain of knowledge and corporate context can be very valuable to an organization and useful in your own career development.

Knowledge is key and to be successful in executive level positions, the engineer needs to gain additional knowledge to equip him or her in these higher-level positions.

Skills

As described by O*NET, "Experience lays the foundation for establishing procedures to work with given knowledge. These procedures are more commonly known as skills." [6]. Skills are used in day-to-day work, and it is important to understand what skills engineers traditionally may possess and those skills that they may need to attain in their career development.

Figure 6 shows a comparison of Skill Areas between disciplines. (See Appendix "Definition of O*NET Knowledge Areas, Skill Areas and Work Styles" for a full definition of the skill areas.)

From Fig. 6, you can see several skills that are listed for general and operating managers and chief executives that are not shown in the engineering and engineering management disciplines. Those skills are listed below.

- Management of Personnel Resources
- Social Perceptiveness
- Management of Financial Resources
- Negotiation

Management of Personnel Resources is important for successful team management and for achieving greater levels of responsibility within a company. You need your team and reporting hierarchy to effectively perform their work to support organizational objectives. In executive positions, you need to be able to direct and leverage your organization's skills to achieve corporate goals. Skills in understanding your team needs, managing

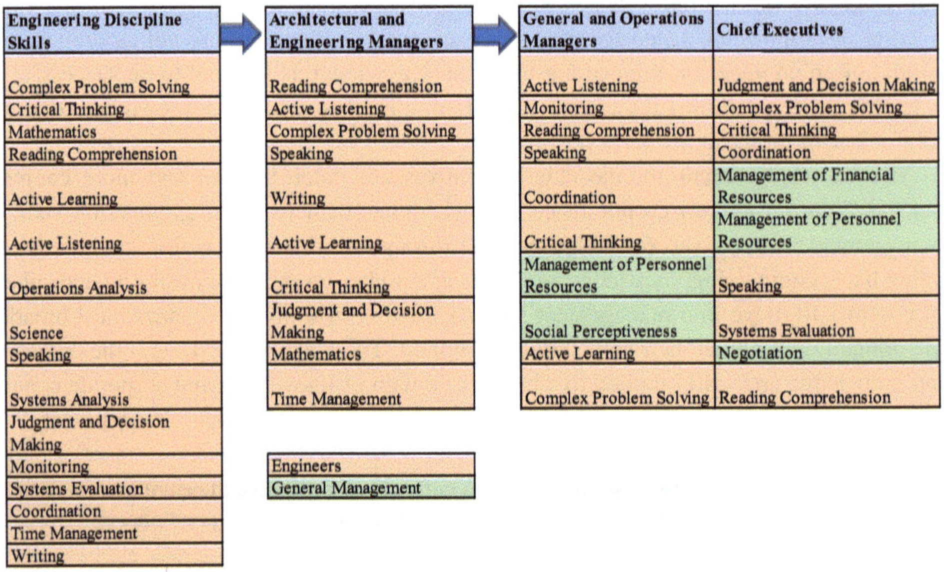

Fig. 6 Comparison of skill areas between disciplines [6]

their development, following appropriate human resource procedures are important for the management of an individual team and an organization.

Social Perceptiveness is a skill that is only listed in the general and operations manager position, however, is critical in understanding your organization, your team, your boss and yourself. This skill area moves you from a more traditional engineer/engineering manager position to leadership positions and provides you with the skills to understand you and your fit within the organization. Social perceptiveness may not inherently be a strong suit of engineers. Engineers like focusing on their technical work. Understanding yourself, your boss and your team is discussed in later chapters of this book. As an engineer, it may not be an area that you really want to think about, but in many ways, it can be career limiting if you don't address its need.

Management of Financial Resources is also critical. This involves understanding you and your team's contribution to the bottom line, critical financial measure that are important to your organization, the ability to justify capital expenditures and successfully managing the resources that you have under your control to support corporate objectives. Budgeting and financial management from a high-level perspective is also discussed later in this book.

Negotiation is a beneficial skill to possess but requires training and experience. You may need to negotiate at a staff engineering level to promote your design options. Moving into mid-level manager positions, you may then need to negotiate the use of various resources to successfully complete your tasks. Negotiation skills can be successfully used when you are involved in projects and project management to ensure that you have available those resources you need to be successful. You may also need to negotiate with vendors about costing and the delivery of products. As you move to roles with more responsibility and span of control, the ability to negotiate is an important skill.

An additional observation about the skills listed above is in System Evaluation. Most engineering skills show some version of System Evaluation, and this skill is shown in the Chief Executive skill set as well. Engineers at a more operational level will have skills in Operational Analysis, System Analysis and System Evaluation. These system's skills are very useful at a base engineering level in integrating your work into a broader operational system. These engineering systems skills can be leveraged and applied at a broader, organizational level by expanding your system's domain view to a broader, more comprehensive perspective of your organization.

Mechanical and chemical engineers may analyze machines, processes, and operations and how they fit in with the overall system operation. Industrial engineers may analyze operations to improve productivity. From an engineering discipline perspective, engineers design and develop individual components but also need to be aware of how those components fit within overall operations. The chief executive is evaluating systems from a broader, more holistic perspective of overall corporate operations. Many of the concepts of system evaluation at a lower level in the organization can translate to a broader view at an organizational level. The skills of system operation and evaluation can be leveraged at higher levels of the organization with a broader breadth of knowledge and holistic context.

There are some interesting observations around the skills of the engineer and engineering manager and Chief Executives in Complex Problem Solving, Critical Thinking and Judgement and Decision Making. Here you see that for the Chief Executive positions, these three skills are at the top of the list of importance. When you look at the skills of the staff level engineer, all three of these skills are important with Complex Problem Solving and Critical Thinking at the top of the skills listed for these engineering disciplines. Engineering disciplines emphasize training in these areas starting from an academic accomplishment perspective. Engineers are required to take complex math, physics and other sciences and to solve complex problems within their academic course of study. Engineers are continually presented with complex problems that require them to research, analyze and provide solutions whether to a professor or to their boss. Individuals with these critical thinking and problem-solving skills tend to succeed. Chief Executives are continually presented with difficult and complex problems that require the same set of skills to successfully lead their organization but apply these skills in a broader context.

There is a critical difference is the domain of knowledge that an engineer must possess and the domain of knowledge that an executive must possess. Engineers at a staff level typically need to possess knowledge and skills in their engineering discipline. Chief Executives must possess knowledge and skills across all company domains to be successful and leverage their team's knowledge to provide them with appropriate information and analyses to make decisions. To expand their career development paths, engineers need

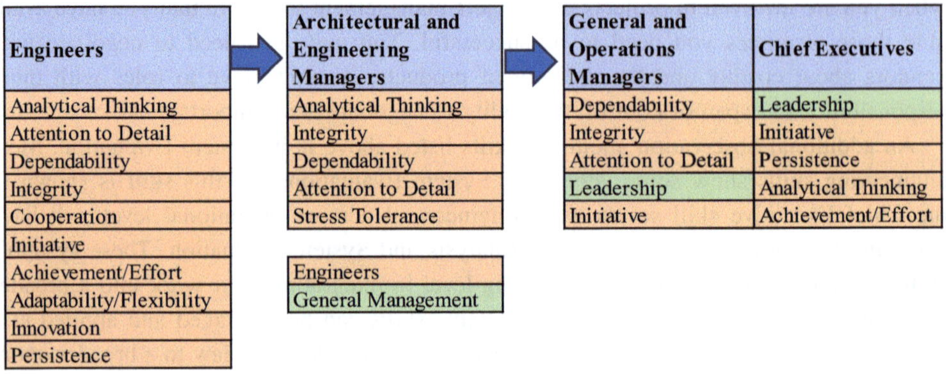

Fig. 7 Comparison of work styles between disciplines [6]

a better understanding of the corporate domain. Engineers' complex problem solving, critical thinking and judgement and decision-making skills will serve them well if they broaden their knowledge base from engineering specific problem solving to corporate wide problem solving. This may be why nearly one third of all CEOs are engineers.

Work Styles

As defined by O*NET, "work styles are personal characteristics that can affect how well someone performs a job" [6]. Work styles represent how you approach your work and the accomplishment of your tasks. Figure 7 shows work styles that engineers possess.

Figure 7 shows a comparison of Work Styles between disciplines. (See Appendix "Definition of O*NET Knowledge Areas, Skill Areas and Work Styles" for a full definition of the work styles.)

The identification of stress tolerance for the engineering manager may be related to additional requirements to meet deadlines and typically, a requirement to handle additional responsibilities as an engineer for their team. The assumption can be that the engineering manager may be given more responsibilities without the clear understanding from management that engineering team management will add complexity to the job.

Analytical thinking is at the top of the list of work styles for both the engineer and the engineering manager. It does not show up in the workstyles for the general and operational manager but does show up in the work styles for an executive. Although not a direct leap, this does lead one to the thought that the analytical thinking that an engineer possesses can translate into the analytical thinking required by an executive.

Another work style listed for the chief executive but to a lesser degree with the engineering disciplines is Achievement/Effort. O*NETS definition of this work style is "Job requires establishing and maintaining personally challenging achievement goals and exerting effort toward mastering tasks." Most engineering disciplines exhibit the Achievement/ Effort workstyle although not listed in the top five workstyles for engineering discipline in the sample, however, this may be expressed from a corporate-wide perspective of accomplishing goals on an organizational level. Executives typically don't mind being in the forefront of activities and may thrive on their organizational accomplishments, which is great for leading the organization. Engineers may be very content in more background positions but still in most cases are driven by accomplishment of their work.

Engineering work styles translate across positions with increased responsibilities except for a focus on leadership. From the article, "Fundamentals of Leadership Has Not Changed," [7] they state success in leadership still fundamentally depends on six timeless areas: vision, strategy, engaging people, focusing on results, leading innovation, and leading yourself." Leadership is a very broad discipline with thousands of books, articles, papers, and seminars being produced every year. It is not the intent of this book to cover all these topics or areas, but to provide some of the elements of leadership that engineers can develop from an engineer's perspective and at a tactical level building on skills that are inherent in engineering study and application.

Career Paths for the Technical Leader

Career Paths for Engineers

The importance of developing talent is crucial to the survival and growth of an organization. Engineers have skills, knowledge and work styles that benefit an organization from a technical execution perspective. Additionally, engineers have some key knowledge and skill characteristics that can translate to broader management and executive positions. Engineers have a choice in the career path that they pursue. Engineers think logically, methodically, and objectively while making decisions based on facts and data, and analyzing and defining technologically feasible solutions. Additionally, engineers understand what motivates other technical professionals and possess expertise that supports high-quality decision-making. The career path of an engineer shows a planned progression to higher-level positions of responsibilities.

Engineers naturally progress to the position of a technical contributor. Initially, an engineer's job is mostly of a technical nature. Engineers then build a reputation for being knowledgeable in technical domains and may be asked to serve as a technical contributor on a team. Technical contributors develop connections with others to build networks, lead teams to empower them to contribute, expand their technical knowledge, exemplify truthfulness and authenticity and sustain their best efforts. The engineer's mindset has been created or evolved through academic training and experience in the workplace.

The engineer's skills afford him or her multiple paths of opportunity. These opportunities can come as a technical contributor, a project manager evolving to a corporate leadership position or progressive management responsibility leading to a corporate

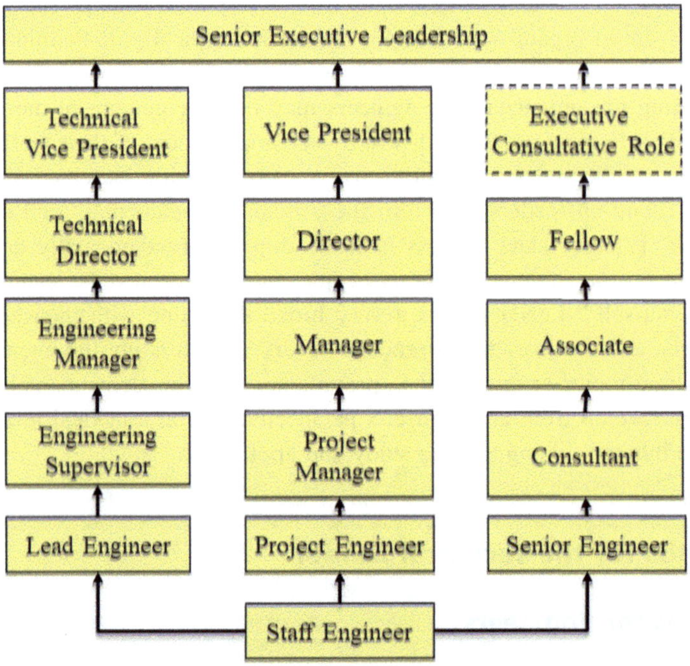

Fig. 8 Example of career path for engineers

leadership position. Figure 8 shows an example of potential examples of progressive responsibility available for an engineer's career development [8, 9].

The engineer's skills afford him or her multiple paths of opportunity. These opportunities can come as a technical contributor, technical manager or a project manager evolving to a corporate leadership position.

After initial technical promotions, the engineer may be asked to make a choice between the technical ladder and the managerial career path. Technical midlevel positions include such titles as senior engineer, consultant, engineering/development associate, and corporate fellow which may then serve as a consultant to executive leadership. On the managerial career path, these positions may include a project manager position leading to more general leadership positions. Engineers can then move from mid-level executive positions to vice president, senior vice president and chief executive positions. Career paths are varied and may cut across these paths. Typical incentives for engineers becoming managers and executives include financial reward, executing authority, acquiring influence, receiving recognition and maintaining relevancy within an organization. Drawbacks include increased workload, stress, maintaining work-family balance, business travel, and others.

Engineers may fail in management positions because they do not possess the knowledge, skills and work styles that general managers and executives possess. This knowledge and these skills can be attained if it is recognized that it is needed, and appropriate training is received. Additional factors that can cause engineers to fail in management positions include personal traits such as lack of political savvy, unable to handle ambiguity and inherent engineering personality traits. From a technological perspective, engineers may be unwilling to take risks and focus on the technological aspect of their position only. Engineers may also not recognize the managerial aspect of their job and have deficiencies in their human relations and management skills. Keep learning, keep reading and keep expanding your skills so that you can mitigate deficiencies that you or others perceive in your ability to hold technical and executive leadership positions in your organization.

Benefits of a Broader View

The remaining chapters in this book addresses components of knowledge, skills and workstyles identified above that will prove helpful in your career development. The field of leadership and management is vast, so the intent is to only address a small component of that field as it relates to engineers and engineering characteristics. The goal is to get engineers to think about additional organizational areas and why they are important. Key management and leadership elements are woven into the various chapters of this book. Core engineering skills like judgement and decision making, complex problem solving, critical thinking and system evaluation can be applied in a broader organizational domain than your specific engineering discipline. Understanding what your organization does and how the various components of your organization interact provides you with a strategic view of your current operation within the context of the broader organization.

Although other knowledge areas listed above will be addressed in this book, the knowledge areas listed below will be highlighted.

- Personnel and Human Resources
- Administrative
- Economics and Accounting
- Sales and Marketing

From a skills perspective, the book will discuss the expansion of core engineering characteristics at a broader level of the organization. Skills such as those listed below will be addressed.

- Judgment and Decision Making
- Complex Problem Solving
- Critical Thinking

- Coordination
- System Evaluation

Additionally, skills that may not be typically taught in core engineering disciplines will be highlighted in the book.

- Management of Personnel Resources
- Social Perceptiveness
- Management of Financial Resources
- Negotiation

From a work styles and overall perspective, the book will address leadership and the translation of core engineering skills to higher levels of responsibility within an organization.

Questions

Below are some questions for you to consider from this chapter.

- How well is the sector of your current employer performing? Do they project growth in the next few years or is the sector declining?
- What knowledge and skills do you feel that you possess and what areas do you feel need further development?
- Begin to think about what you really want from your career. Are you looking to move into middle or upper-level management, or do you want to pursue a technical leadership position? Do your near-term and long-term goals follow are similar path?
- Review Knowledge, Skills and Work Style information provided. Do you see yourself in this information?
- Have you thought through your career path? Do you want to be a technical expert, a technical manager, a general manager or an executive?
- Do you need more knowledge and skills to move you forward in your career?

References

1. Aquino, Judith, *Surprising Facts About CEOs*, March 23, 2011, Business Insider, Surprising Facts About CEOs (businessinsider.com)
2. Rottmann, C., Sacks, R. Reeve, D., Engineering leadership: Grounding leadership theory in engineers' professional identities, Leadership 2015, Vol. 11(3) 351–373, Sage Publication, Newbury Park, CA

3. North American Industry Classification System, United States, 2022, Executive Office of the President, Office of Management and Budget
4. Colato, Javier, and Ice, Lindsey, Industry and occupational employment projections overview and highlights, 2022–32, October 2023, Monthly Labor Review, Bureau of Labor Statistics, Industry and occupational employment projections overview and highlights, 2022–32
5. Ice, Lindsey, A Look at Projected Employment in Professional, Scientific, and Technical Services, 2021–31, March 2023, Bureau of Labor Statistics, A Look at Projected Employment in Professional, Scientific, and Technical Services, 2021–31: Spotlight on Statistics: U.S. Bureau of Labor Statistics (bls.gov)
6. O*NET Resource Center (onetcenter.org) This page includes information from the O*NET 28.1 Database by the U.S. Department of Labor, Employment and Training Administration (USDOL/ETA). Used under the CC BY 4.0 license. O*NET® is a trademark of USDOL/ETA
7. Ashkenas, R., Manville, B., The Fundamentals of Leadership Still Haven't Changed, November 7, 2018, https://hbr.org
8. Cabanes, B., Galey, Ph., Le Masson, P., Weil, B., Technical Staff Management for Radical Innovation in Science-based Organizations: a New Framework Based on Design Theory July 2016 Conference: R&D Management Conference, Jul 2016, Cambridge, United Kingdom, Cambridge, United Kingdom
9. Rice University, George R. Brown School of Engineering, What Does an Engineering Manager Do, February 20, 2023, What Does An Engineering Manager Do? | MEML@Rice

Understand the Big Picture

Abstract

Understanding your organizational environment is important to your success. You should understand organizational functions outside of typical engineering disciplines and why these organizations are important and impactful to you as an engineer. If you haven't assessed your structure and fit within your organization, you should. It is important to know the hierarchical relationships that exist in your organization and how you can best interact with individuals outside of your immediate organization. Do you have a good understanding of yourself and your work styles? Personality assessments, approaches to organizational interaction and career planning can provide you with insights. Have you assessed your boss? Your boss has work styles and personal characteristics that you should be attuned to best interact with your boss and maximize your career opportunities. Knowledge, skills and workstyles identified in the first chapter that are addressed in this chapter include the following:

Knowledge:

- Personnel and Human Resources
- Economics and Accounting
- Sales and Marketing

Expansion of core engineering characteristics at a broader organizational level:

- Critical Thinking
- Coordination
- System Evaluation

D. Cassone, *Leadership Concepts for the Engineering Mindset*, Synthesis Lectures on Engineering, Science, and Technology, https://doi.org/10.1007/978-3-031-67798-4_2

Skills:

- Management of Personnel Resources
- Social Perceptiveness
- Management of Financial Resources

Leveraging Core Engineering Skills from a Broader Perspective

The information in the previous chapter shows that engineers have many of the core skills to make them successful not only in their engineering disciplines, but also successful in overall corporate leadership positions. Key to their continued success is gaining a broader domain of understanding of the organization and its functions. Engineering disciplines specifically teach and apply knowledge and skills in their individual disciplines and professions. Expanding your domain of knowledge provides you with additional content to apply those core engineering skills. The information in this chapter contains an overview of some key functions and information to provide you with broader areas to utilize those skills. The functions are presented from the perspective of why engineers should learn more about these functions and how they can benefit an engineer. Understanding how a corporation functions from a bigger-picture perspective will provide you with additional context to be successful in your technical and leadership development activities.

Information in this chapter is not meant to by any means be exhaustive. Entire fields of study and experts exist and should be sought after and studied to expand your knowledge of corporate operations. You need, however, to know that you should learn and leverage this information in your day-to-day work to apply your engineering skills to a broader knowledge domain.

Corporate Functions and the Big Picture

You may not have needed to interface with individuals outside of your engineering domain at this point in your career. On the other hand, you may currently be interfacing with individuals and teams outside of your domain to accomplish your current tasks. It is important to have a general understanding of typical areas within a corporation and to understand how these functions may impact you.

In your current engineering job, you may not interface with individuals outside of your organization. You should understand these functions and how they impact you.

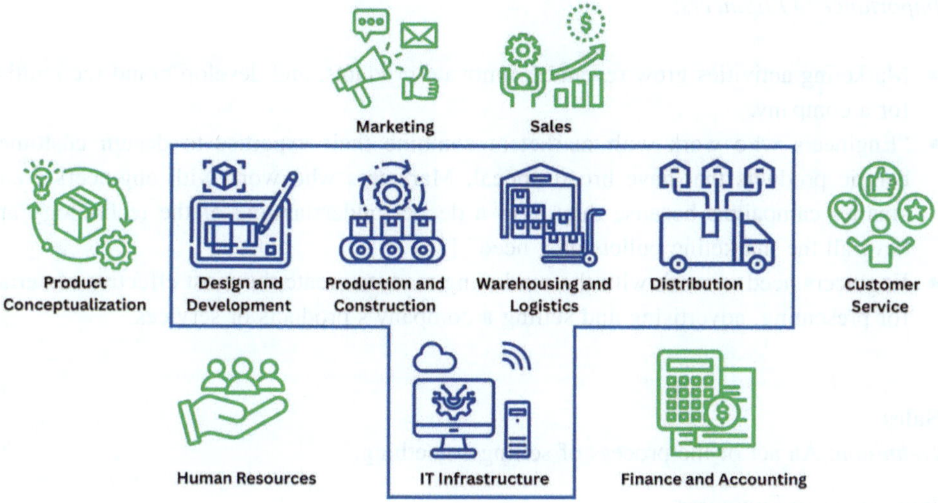

Fig. 1 Typical organizational functions

When you interface with organizations outside of your function, it is equally important to know and build your internal network of individuals that you can work with to accomplish your goals. Figure 1 provides a view of overall corporate operations. Highlighted in blue are those functions that various engineering disciplines may traditionally fall within engineering disciplines. Those functions include design and development, production and construction, warehousing, logistics and distribution and information technology infrastructure. Each organization is different, but typically, some versions of these functions serve as heart of overall operational tasks within an organization.

Areas that engineers may not traditionally begin their career are in marketing, sales, customer service, human resources, finance and accounting and product conceptualization. Engineers know that these functions exist, however, they may not be aware of the function's impact on their work. A brief description of these functions is given below and how each of these areas can impact your work as an engineer. Entire fields of study and much more comprehensive definitions functions exist, however, the brief definitions listed below are from Oxford Learner's Dictionary [1]. The focus below is on why an engineer should know and understand each of these functions, not a comprehensive description of the field.

Marketing

Definition: The activity of presenting, advertising, and selling a company's products or services in the best possible way.

Importance to Engineers:

- Marketing activities grow revenues, innovate products, and develop brand recognition for a company.
- "Engineers who work with marketers combine their expertise to design customer-centric products that have broad appeal. Marketers who work with engineers create smarter campaigns because they have a deeper understanding of the technology and have all the marketing bullets they need" [2].
- Engineers need to work with the marketing team to create the most effective materials for presenting, advertising and selling a company's products or services.

Sales
Definition: An act or the process of selling something.

Importance to Engineers:

- Sales builds a bridge between marketing and engineering. The sales force needs to understand the product or service and communicate its value to potential customers. Engineers can help with that communication.
- The sales team also provides valuable feedback to the engineering and development team. Sales teams identify design and operational flaws, new product features, and customer service needs for the successful sales of a product. This information is important to engineers to refine and approve their designs.

Customer Service
Definition: The help and advice that a company gives people who buy or use its products or services.

Importance to Engineers:

- Complex products may require detailed knowledge in customer service functions.
- A full understanding of product features, operational requirements, product use cases and product limitations is needed. Engineers can assist in providing this information.
- Customer service representatives are the front-line for product use and provide valuable feedback to the engineering team.

Finance and Accounting
Definition: Money used to run a business, an activity or a project.

Importance to Engineers:

- Finance and accounting functions assess concrete value and costs across all areas of operations.
- Engineers should understand the cost structure of their products and designs and look for ways to optimize the cost structure.
- Engineers should be familiar with business case development to support product development and operational improvements.
- Costs drive budgets and senior leadership is accountable to stockholders and the board of directors for financial performance.
- Prudent financial management of company resources is a cornerstone of business success and longevity.

Human Resources
Definition: People's skills and abilities, seen as something a company, an organization, etc. can make use of.

Importance to Engineers:

- "Engineers learn leadership and management skills while working learning "soft skills the hard way"" [3].
- Engineers can be overlooked in senior management positions due to their deficits in human resource management. Engineers can take an active role in enhancing their personal development.
- Additionally, as engineers move into management, they need to understand how human resources operates within their company and what rules, processes and procedures are required by the company to manage team members.
- Lack of understanding of these requirements can be a detriment to them and leave their company open to liabilities by not appropriately managing their team.

Product Conceptualization
Definition: To form an idea of something [a product] in your mind.

Importance to Engineers:

- Product conceptualization and new product development provide an exciting opportunity for engineers. Engineers can be "heads down" in designing and developing products. New product development provides engineers with the ability to innovate and provide companies with new product offerings.

- Product conceptualization typically requires a cross-functional team to assess the product viability, so engineers should understand the broader decision-making team responsible for new product launches.
- Engineers that demonstrate the ability to innovate and work across corporate functions are provided opportunities to contribute to this corporate function.

There are elements of the descriptions and indicated importance that may seem obvious to many. Engineering curriculums may have changed some to include some of these functions. Initially, in the engineer's career you may have limited interaction with these other functions, however, as your career development progresses, these functions could become as important, if not more, than your engineering discipline training.

More traditional business functions may become as important, if not more important, than your engineering discipline training as your career progresses.

The intention of this section is to provide engineers with the reason why they should know more about these functions. Many of the corporate functions listed above are not part of an engineering academic discipline and engineers may view these functions as foreign and unnecessary to their career development. To enhance an engineer's domain of knowledge, it is beneficial to learn more about these functions and think through how they can leverage this knowledge in their careers.

Your Organization's Fit Within the Organization Structure

Where you currently fit within an organizational structure is part of a broader view of overall corporate operations. Many different types of organizational structures exist and only a limited few of them are discussed in this section. Engineers can be highly focused on delivering their technical work, but it is important to understand where you sit in the organizational structure and your organizational chain of command. This helps you to know who and how you can interact with individuals within your organization.

A traditional organizational structure is shown in Fig. 2. The traditional organizational structure is organized by functions where senior level management reports to the chief executive functions representing their functional responsibilities. This structure represents a chain of command where lower-level individuals report to those with increasing levels of responsibility and span of control. You should know your reporting chain of command within the organization as well as cross-functional peers within your organization to better support your work efforts.

A "flat" or horizontal organizational structure contains less of the hierarchical chain of management and provides greater access to individuals and decision makers. A horizontal organizational structure may be useful for companies that are smaller in size and

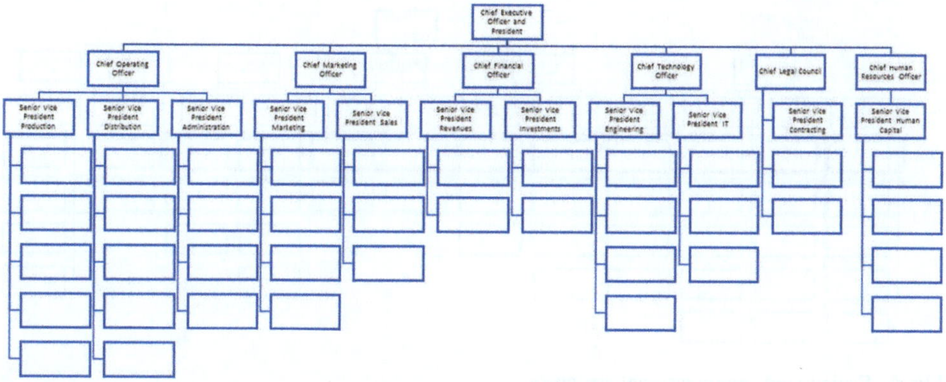

Fig. 2 Traditional organizational structure

require heavy cross-functional interactions. Large companies may not have the ability to operate from an overly flat organizational structure due to the number of employees and the specialization of work required to support the organization. Horizontal org structures are typically designed to enhance the accessibility of individuals within an organization. Engineers may more freely interact with other functions with this organizational structure because there is less formality and bureaucracy in the organization. Figure 3 shows an example of a flat organizational structure.

Another type of organizational structure is a heavily projectized or matrixed organizational structure. This type of structure has a heavy focus on projects and project management activities. With this structure, functional hierarchies are still maintained, but teams are led by project and program managers who interface cross-functionally to engage resources and deliver work. Engineering firms may use this type of structure when engaging in larger, more cross-functional projects. Figure 4 shows an example of a projectized organizational structure.

Corporations can have a myriad of organizational structures. The key point for you to understand is that you should know how your organization is structured and understand

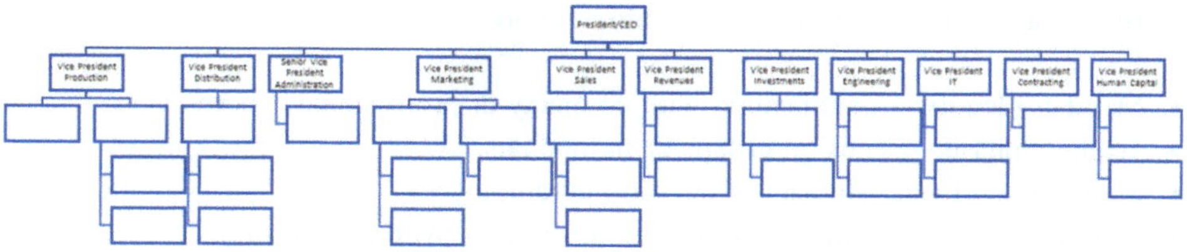

Fig. 3 Flat organizational structure

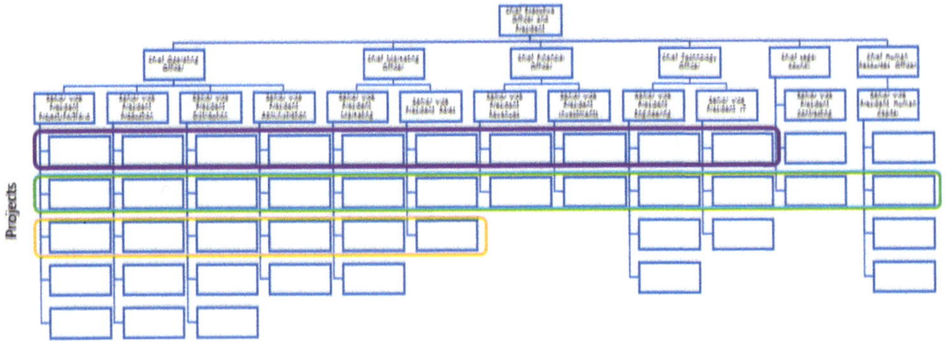

Fig. 4 Projectized organizational structure

the reporting relationships within your organization. This will assist you in understanding how best to interact with management, peers and employees.

As an engineer, you should understand how your organization is structured and how best to interact with individuals in your organization.

Corporate Culture

Corporate culture is a collection of beliefs, values, customs, traditions, and practices shared by an organization's members [4]. Cultural expectations create norms of acceptable behavior used to perform work within your organization. It refers to "the practice or method by which things get done" in a social environment. It encompasses the values, practices and procedures based on policies of the organization. In sum, it is individuals as a team agreeing upon the common values that bind them [5].

It is important to spend time to understand what these behavioral expectations are in your organization. Your message can be lost in the delivery of it. Understanding how to best present information and interact with individuals at various levels of the organization significantly impacts the translation of the information you are communicating.

Below are some dimensions of corporate culture.

- Member identity—Identification with company or professional expertise.
- Team emphasis—Focus on groups or individuals.
- Management focus—Impact of decision on employees.
- Unit integration—Coordination or interdependence with organization.
- Control—Effect of rules, policies, and supervision on behavior.
- Risk tolerance—Encouragement of risk taking and innovation.
- Reward criteria—Rewards for accomplishments and punishment for failures.

- Conflict tolerance—Acceptance of conflicts and criticisms.
- Methods—Focus on outcomes versus processes.
- Openness—Adaptability of the environment.

The impact of culture can be seen in the following.

- Provides a sense of identity for its members.
- Clarifies authority relationships with regards to positions of authority and appropriate respect of individuals in those positions.
- Creates a social order.
- Provides a basis for predictable behavior that promotes efficiency in operating norms.
- Identifies subcultures that may exist in various divisions or departments within an organization.

Engineers, in particular, may believe that the factual work that they do speaks for itself. A design may be the most efficient approach to solving a problem, however, if delivering the information about the design is not done within the normal operating environment, the message can be lost. Although the work is correct and logical, if not presented within the operating characteristics of an organization, it may not be accepted. Watching, learning, and practicing acceptable ways to present information and interact with superiors and peers is important to enhance the acceptance of your ideas and results. Doing this incorrectly can be career limiting.

Watching, learning, and practicing acceptable ways to present information and interact with superiors and peers is important to enhance the acceptance of your ideas and results. Doing this incorrectly can be career limiting.

Subcultures can be an interesting phenomenon within an organization. For example, a marketing organization may have more latitude to make decisions on marketing campaigns with more time and collaborative engagement with the organization. A logistics-based organization may not have that luxury. Picking, packing and shipping products to meet customer expectations may drive the need to have a "get it done" attitude within an organization. Individuals in a logistics organization may not have the time to adhere to an organizational corporate culture because they need to get products out of the warehouse to the customer.

As an engineer, it is good to be aware of your corporate culture. In doing so, you can decide whether you want to integrate and assimilate with the culture or find another organization that better fits your operating style. Take the time to assess how you fit within your organizational culture. Your acceptance of the culture is important in assessing your longevity at an organization and your personal job satisfaction.

You and Your Boss

It is important to understand yourself as well as your boss. To operate in cohesive roles, you should understand your key personality traits and then look at your boss and assess him from that perspective as well. This will help you understand how best to interact in your immediate reporting responsibilities in an organization.

Numerous personality profile tests exist, such as Meyer-Briggs, DiSC, Test Color and many others that can help to provide you insights into your personality and operating style [6]. These tests are readily available online and can be helpful in providing you with insights into yourself and how you operate in an organization.

Some elements of how your personality may translate into the workplace and your success include technical acuity, personal acuity, career aspirations, problem solving approach, solution delivery timing and your personal life and work-life balance objectives.

Assessing Yourself

Below are some areas that you may consider when assessing yourself.

Technical acuity
Questions: Are you a new staff engineer with a high learning curve in your organization? Are you viewed as an expert in your organization? Do you continually engage in learning new technical skills? Are you able to implement your knowledge in your organization? Do you have a realistic view of your skills and contributions to an organization?

Discussion: Of all the workplace characteristics, understanding your technical acuity may be the easiest of them all. You may recognize individuals within your organization that are highly skilled in your technical discipline, or you may be that individual that mentor's new engineers in their work. Keep looking and learning in your field to stay on top of your professional discipline.

Personal acuity
Questions: How would you rate your emotional intelligence? How are interactions with your peers? Do you know what your personal strengths and weaknesses are? Do you find yourself in cooperative or conflicting situations? How do you handle conflict? What would others say about you? Are you viewed as a team player? Do you interact better with your peers than with your boss and upper management?

Discussion: It is important to have a good understanding of yourself and your work style within an organization. Taking a personal assessment enables you to gain an understanding of areas that you excel in personal interactions and areas that you may need to work at improving. How you interact in an organization can lead to success or become career limiting. You may also want to look at the corporate culture and see if it is truly a fit for you. In one organization you may be viewed as combative, while in a different organization it may be acceptable behavior. It is important to be self-aware and manage yourself within the context of the work environment.

Career aspirations

Questions: What are your near-term and long-term career goals? Do you see yourself as a staff engineer, a technical leader, moving into engineering management positions or embracing a broader, cross-functional leadership positions? Is your current position career limiting? Does your current position provide advancement opportunities? Is your current position moving you towards your long-term goals? Do you need additional training or education to achieve your long-term goals?

Discussion: You should take a careful look at your current position and what you may need to learn or develop to reach your long-term goals. Plot out a path to achieve your long-term goals. Determine what education, training and opportunities you may need to achieve those goals. Make it an explicit plan instead of allowing circumstances to dictate your future. Don't be afraid to change your goals or take on new opportunities that may not be part of your original plan. They may lead you to new and exciting career changes that you may not have thought about on your own. Taking the more difficult road may challenge you and develop your skills in ways that you had not planned.

Problem-solving approach

Questions: Do you like to solve problems alone or in team situations? Are you data-centric or are you consensus-centric in your problem solving and decision making? Do you rely on other technical and non-technical resources to guide your solutions? Have you considered the big picture needs of the organization? Do you apply your engineering critical-thinking skills to only technical problems or to those problems that are non-technical?

Discussion: Understand how you like to solve problems. You may want to work independently to develop solutions. You may want to gain consensus in your problem-solving approach. You may also want to take a hybrid approach and develop solutions in an iterative manner. Ensure that you are developing solutions that align with corporate objectives. Developing a technical solution that is difficult to manufacture or implement may be exciting scientifically, however, may not be practical in implementation. Problem solving should involve your team and your boss, also, to gather input and ensure that you are arriving at the best possible solution. Others may see things that you do not and consulting and respecting their opinions may lead to a better overall solution.

Solution delivery timing
Questions: Do you deliver your work on time, or do you procrastinate? Are you able to set priorities that are consistent with your boss's priorities? Can team members rely on you to deliver your work? Are you usually the first to deliver or the last to deliver work that will contribute to the team's success?

Discussion: It is important to prioritize your work and ensure that your priorities are consistent with your boss's priorities. Typically, your work will impact the ability for others to complete the work for the team. When at all possible, deliver early. This allows for time to review and revise work and gather valuable feedback from your team members. Don't be the employee that causes delays.

Personal life/work-life balance objectives
Questions: What are your personal goals? What hobbies and activities do you have outside of work? Do you have family responsibilities? What other aspects of your life give you personal satisfaction? Do you have responsibilities outside of work that impact your work life? Are there career choices that you may make now that will differ at another point in time?

Discussion: It is important to be well rounded and have areas of interest outside of work. This helps give you personal satisfaction and purpose that you may or may not find in your work-life. When possible, make intentional decisions about your work-life balance. Make those choices explicitly to avoid regrets.

Explicitly assess your technical acuity, personal acuity, career aspirations, problem solving approach, solution delivery timing and your personal life and work-life balance objectives to make career choices and avoid regrets.

Assessing Your Boss

It is important to understand your boss and your boss's perspective. Many times, an employee does not take the time to think about their boss strengths as well as weaknesses, understand what motivates their boss, assess their boss's position within the organization, their boss's background and their boss's relationship with their chain of command in the organization. Assessing your boss provides meaningful insights into your interactions with him or her and your interactions with the organization, overall. Key points from Hill, Linda A., and Kent Lineback, Being the Boss: The 3 Imperatives for Becoming a Great Leader, are discussed below [7].

Managing up. Managing up is important because your boss has a pivotal role in your success or failure in the organization. By better understanding your boss and his or her position in the organization, you can leverage your boss's influence in the organization with regards to information, resources, advice, organizational knowledge and access to higher levels of management. This leverage can be beneficial in the work that you are doing and can be helpful in your career development path.

Fostering your relationship. Your boss is someone who can have a direct impact on your career, so fostering your relationship with your boss can be very important to your success. Assess whether your boss views you as a partner where you need each other to succeed or is your boss an evaluator and judge. The best relationship can be one of mutual dependence on success, however, you may have a highly authoritative boss that does not approach management from that perspective. Respecting your reporting relationship is important independent of the type of management style of your boss.

Past boss baggage. Do you see past bosses in your current boss? Many times, experiences from previous bosses translate to your current relationship, however, your past boss may be a completely different person. Avoid carrying this baggage with you to a new position and new reporting relationship.

Meeting expectations. Take stock of your current relationship with your boss. Do you operate as partners? Do you have constructive discussion about work? Are you able nudge your boss in areas that you feel are important? Importantly, to be respected as an employee and to be considered as a positive contributor, think about your own performance and how your performance contributes to what your boss needs to accomplish. Meeting expectations is important to developing trust with your boss. This shows that you are going to provide your boss with the information and support that they need to succeed.

Seeing your boss as a person. An interesting question about your boss is whether you see him or her as a person, or only as an individual with a position of authority over you. Do you know your boss's aspirations, frustrations, fears, strengths, and weaknesses? What is your boss's background and training and are they well suited for his or her current position? What is your boss's personal life, family, social organizations or hobbies? These questions do not imply that your boss should be included in your weekly Sunday barbeques and social events, but it is worth noting these additional aspects of your boss to help you view his interactions in a professional environment and how some of his or her personal traits may influence your interactions at work.

Management style. When you think about your boss's management style, what do you observe? Are there obvious strengths and weaknesses? Has your boss been passed over for a promotion? How does your boss approach decision-making? Are decisions made from a data-centric perspective and does your boss make decisions independently or by group consensus? Can you identify your boss's strengths? Many times, employees focus on their boss's weaknesses, but it is also important to focus on their strengths.

Spending time assessing and understanding your boss is helpful in your work and reporting relationship. You may inherently do this assessment, or you may assess pieces and parts of your work with your boss, especially as they impact your work and career.

Does your boss know your career development goals and aspirations? This helps him or her with your coaching and development. Overall, it is helpful to take some time to think about your work relationship and operating characteristics of your reporting structure to gather insight and develop strategies to assist in accomplishing your work and improving your work environment.

Spend time assessing and understanding your boss. Assessing your boss provides meaningful insights into your interactions with him or her and your interactions with the organization.

Questions

Below are some questions from this chapter for you to consider.

- Why is it important for you to have a broader understanding of your organization?
- Why should functional areas like sales and marketing be important to engineers?
- What is your company's organizational structure and how does it impact your accessibility to senior level management?
- What is your company's culture and how does it impact you in your day-to-day work?
- How well do you know yourself, your work style and interactions with peers, subordinates and management?
- How well do you know your boss, his or her management style and how best to interact with him?

References

1. © 2023 Oxford University Press, Learner's Dictionary, Oxford Learner's Dictionary of Academic English at Oxford Learner's Dictionaries (oxfordlearnersdictionaries.com)
2. Crawford, M., Engineers Need Marketing Skills – ASME, American Society of Mechanical Engineers, April 6, 2023, Engineers Need Marketing Skills - ASME
3. Kumar, S. Hsiao, J. K., Engineers Learn "Soft Skills the Hard Way": Planting a Seed of Leadership in Engineering Classes, Publication: Leadership and Management in Engineering Volume 7, Issue 1, January 1, 2007, https://doi.org/10.1061/(ASCE)1532-6748(2007)7:1(18)
4. Tarver, E., Corporate Culture Definition, Characteristics, and Importance, April 25, 2023, Investopedia.com, Corporate Culture Definition, Characteristics, and Importance (investopedia.com)

5. Rebello, K., The Impact of Culture in the Workplace, Forbes.com, December 1, 2021, The Impact Of Culture In The Workplace (forbes.com)
6. Robinson, A., Personality Tests: Quick, Fun and Free for Work in 2023, Teambuilding.com, December 12, 2023, Personality Tests: Quick, Fun, & Free for Work in 2023 (teambuilding.com)
7. Hill, Linda A., and Kent Lineback, Being the Boss: The 3 Imperatives for Becoming a Great Leader, Harvard Business Review Press, 2011. Being the Boss: The 3 Imperatives for Becoming a Great Leader

Defining What Your Team Does

Abstract

Developing task profiles and assigning team responsibilities provides a clear picture of what your organization does and who is responsible for the work. The summary of a study developed for a large government agency is discussed and a simple method for developing task profiles and work responsibilities is provided. Task profiles of your organization are beneficial in developing organizational assignments, identifying "single points of failure," providing a team building activity to assist your employees in understanding the team members' responsibilities, and providing quantifiable information that can be communicated to upper management and assist with restructuring and workforce reduction activities. Knowledge, skills and workstyles identified in the first chapter addressed in this chapter include the following:

Knowledge:

- Personnel and Human Resources
- Administrative

Expansion of core engineering characteristics at a broader organizational level:

- Critical Thinking
- Coordination
- System Evaluation

Skills:

- Management of Personnel Resources
- Negotiation

D. Cassone, *Leadership Concepts for the Engineering Mindset*, Synthesis Lectures on Engineering, Science, and Technology, https://doi.org/10.1007/978-3-031-67798-4_3

"Tell Me What These People Do!"

The author was serving as a technical expert within a Fortune 100 company and then was put in charge of a team of fourteen individuals. Based on prior career responsibilities, the author preferred organizational assignments as a technical expert, but in an organization where layoffs occurred on more than a yearly basis, the author accepted the position to manage a primarily non-technical team. The group had a somewhat nebulous name, the order management team, and the director of the logistics inherited the team under his management domain during a corporate reorganization. In the first one-on-one with the director, he spent a small amount of time with management niceties, but then very directedly said, "Tell me what these people do!" Luckily, experience defining tasks and workload while consulting could be leveraged to address the demands of the new boss and the information was provided to him.

This chapter addresses some tactical ways to best understand what the individuals in your organization actually do. Initially, a consulting study will be discussed and then a simplified example will be discussed for developing a quick view of your team, their work functions and the coverage and backup you have for organizational tasks.

White-Collar Workforce Management Study

This section provides a generalized summary of a white-collar manpower planning and workforce management study. The study is presented because many times, engineers will hold positions managing professional employees serving in a corporate environment. This study was conducted for a federal agency and specifically focused on 3,700 professional employees, their workload, and their workload drivers. A model was developed to represent the work performed by the staff and then use workload drivers to project the total staff requirements. The original model development process was a multi-year project and took a significant amount of effort. Concepts from this model have been applied in a simplified approach to which may be useful in managing your organizational team. The broader application of this study is provided for exposure to a larger organizational implementation.

Manpower Study Summary

This study involved developing an innovative approach to manpower planning. The study incorporated the traditional Industrial Engineering approach of work measurement in a nontraditional way. The approach eliminates the need to measure each task time individually by developing a task time profile for a grouping of similar tasks and simply identifying the type and frequency of the various group tasks. The study integrated

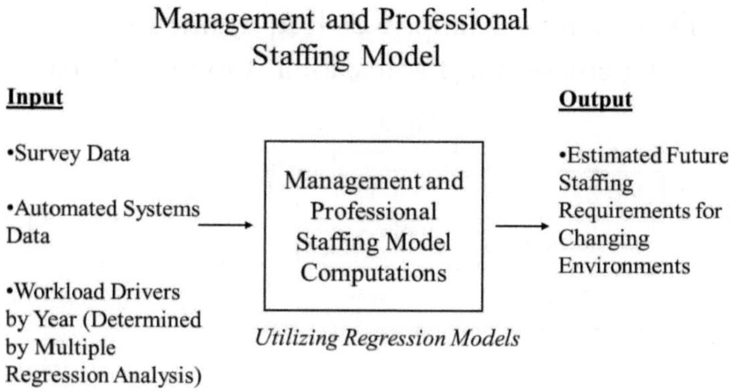

Fig. 1 Holistic approach to manpower planning

management staffing requirements at multiple levels of the organization into a holistic model to plan organizational requirements at an organizational level. (See Fig. 1. Holistic Approach to Manpower Planning.)

As part of the process, task requirement drivers were developed which were used to forecast and monitor workload changes as organizations grow and their product mix or services changes. Once the model was developed, it was dynamic by its nature and could easily adjust the manpower requirements as the workload changes. It was set up so that it monitors key metrics (drivers) that were used to predict workload changes. Utilizing a task data bases, you could quickly change the manpower requirements as the work changes. The task databases contained the average time for the tasks and task groupings and the associated variability for both the demand and production services. A level of risk is established for not filling all of the demand and the associated manpower level for this workload is immediately determined.

The process involved in the development and execution of the Dynamic Manpower Planning Approach is shown in Fig. 2. Product demand metrics or workload drivers were identified for the various levels of the organization. Specific tasks were identified at the various levels of the organization (headquarters, plant operations, etc.) and by the functions (production, management, marketing, etc.) performed within the organization. These tasks were grouped to represent different functions performed within the organization. Data such as the frequency of task, the type of tasks and the time required for each task are gathered into a Task Data Base. Included were the average task times and the variability associated with these tasks and task groupings.

Specific tasks were identified for each of the headquarters' activities, production operations and support requirements. Relationships were established between manpower requirements and demand or workload drivers so that forecast models can be used to determine changes in manpower requirements. These relationships were used to determine the total organizational manpower requirements and skill mix for a given workload

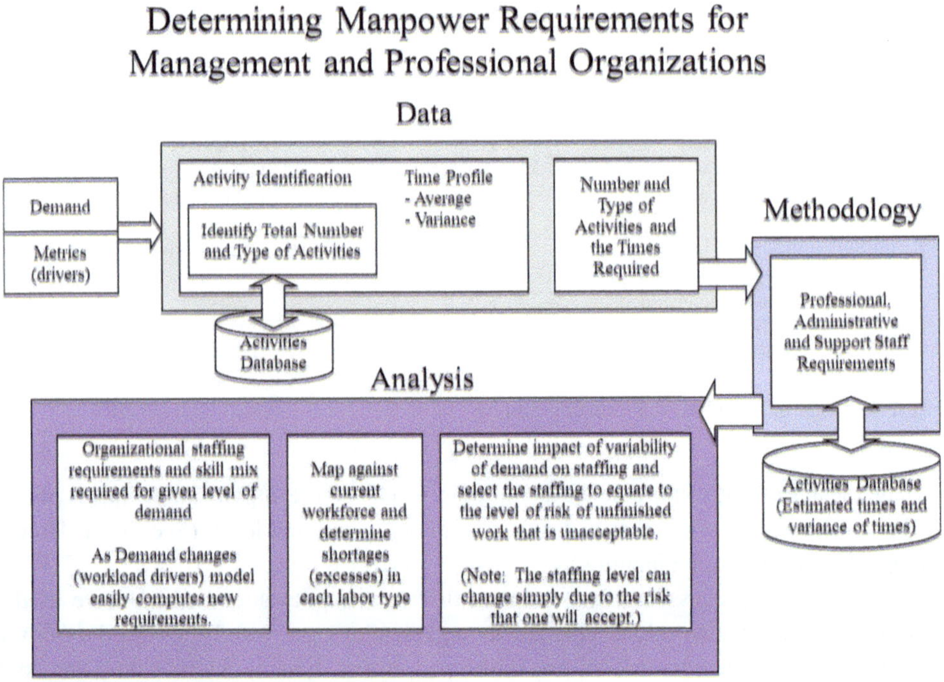

Fig. 2 Dynamic manpower planning approach

or level of demand and then can be compared against the current workforce to determine shortages and excesses by labor types.

The variability associated with demand and production was used to determine a level of risk associated with a given staffing level. Acceptable staffing levels could then be related to various levels of production and demand and manpower decisions are then made based on the level of risk that is acceptable for not completing the required work in a timely manner.

The dynamic model quickly adjusted manpower requirements for changes in the demand, the product mix, and the workload changes. "What if", analysis could be performed to study the impact of changes for various level risks and product demand. This manpower planning approach required minimal levels of work measurement data and provided an ability to accurately determine organizational manpower requirements in a changing work environment. (Section reprinted in part from [1].)

Analytical methods applied to organizational manpower planning are useful in designing, analyzing and optimizing your workforce.

Quantifying Tasks and Task Profiles

The model described above required a significant amount of effort to develop and implement. Once implemented, it was relatively easy to maintain and update. Many organizations may not have the appetite to embark on a holistic organizational model such as the one described above. The experience from the development of the holistic organizational model, however, has driven the simplified use of a key component of the model. This involved simply categorizing and listing the tasks that everyone in your organization performs and ensuring that there is both primary and secondary coverage of those tasks.

Although a very simple process, often managers and executives don't take the time to develop an overall view of organizational responsibilities across their team, assigned responsibilities and back-up personnel. The process to develop your organization's task matrix is relatively simple. The key steps in the process are listed below and shown in Fig. 3.

1. Have each employee provide you with the ten key tasks that they perform.
2. Assemble the information provided by your team into one matrix, with tasks listed horizontally and individuals listed on each of the columns.
3. Perform a preliminary review of the information to combine like tasks or remove duplicates.
4. Call a meeting with your entire team and discuss each task.
5. Make an assignment of "Primary" and "Secondary" for each task as you discuss each task with your team.
6. Review the task list and assignments on a periodic basis to see if your organization's work composition has changed and make modifications if your workforce has changed.

It may appear that this process is like the development of a Responsible—Accountable—Consulted—Informed (RACI) matrix that signs individuals as R—Responsible, A—Accountable, C—Consulted and I—Informed, and it does have some of the similar elements. On numerous occasions, RACI matrices that are developed still lead to confusion. Questions asked in the development of a RACI matrix always revolve around "What does Accountable mean?" and "What does Responsible mean?" The task list development process described above simplifies the process. Individuals know whether they are primarily responsible for completing the task and whether they are secondary, or the backup person, that also completes the task. You as the manager or executive are ultimately accountable for your team's completion of their work, so having a clear understanding of what tasks your team perform, who is primarily responsible for the task and the back-up person assigned to the task is critical.

Tasks	Bob	Jim	Lisa	June	Alex
Task 1	Primary		Secondary		
Task 2		Primary		Secondary	
Task 3	Secondary		Primary		
Task 4	Primary				Secondary
Task 5			Primary		Secondary
Task 6		Secondary			Primary
Task 7	Primary		Secondary		
Task 8		Primary		Secondary	
Task 9	Secondary		Primary		
Task 10	Primary				Secondary
Task 11			Primary		Secondary
Task 12		Secondary			Primary
Task 13		Secondary		Primary	
Task 14		Primary	Secondary		
Task 15					Primary
Task 16			Secondary		Primary
Task 17	Secondary		Primary		
Task 18		Secondary		Primary	
Task 19				Secondary	Primary
Task 20	Primary	Secondary			

Fig. 3 Task assignment matrix

You as the manager or executive are ultimately accountable for your team's completion of their work, so having a clear understanding of what tasks your team perform, who is primarily responsible for the task and the back-up person assigned to the task is critical.

Using Task Profiles for Organizational Assignments, Restructuring and Reductions

The development of the task list for your team helps you manage your team in several ways.

- Provides a clear picture of the work that is performed by your organization.
- Provides a clear assignment of tasks to individuals in your organization.
- Identifies "single points of failure" in your organization when there is no "Secondary" or backup individual to perform your work.
- Enables the entire team to see the responsibilities of their other team members.
- Provides you with information to assess the work that everyone is doing and rebalance work across your team.
- Fosters team building by assisting team members in understanding the work performed by the overall organization.

Developing your team's task list and assignments can have broader organizational value. Companies go through restructuring periodically, and having the information about what your team does and who is doing it can help you and upper management make decisions in the restructuring effort. This information can also be helpful when workforce reductions are required. Having your team's task profiles available, you can assess what work can and cannot be supported with a reduced workforce and assist you in making work reassignments. Conducting this exercise prior to any necessary reorganizations or reductions will help you gain a more objective, and less protective, view of the work being performed by your organization.

A simple exercise of developing task lists across your employees and reviewing primary and secondary responsibilities with your team can assist you in workload balancing and organizational restructuring efforts.

Your organization may not be using this approach to developing task profiles and assignments, but your development of this information can be an example and may prompt others to develop the same information for their team. Again, it is best to develop this information before your broader organization may need it, so that you can get clear and objective information from your team.

The author has specifically done this exercise with multiple teams and seen this approach adopted at higher levels of the organization. Years after leaving a position, previous peers have indicated that this information was still being adapted and used.

Team members like to see that you know the work that they are doing and that you are assessing workload and responsibilities. You seem much less aloof as a manager or executive if you clearly know what your team members are doing and have looked at the workload across your team. This information is also helpful in developing performance appraisals for your team and provides an objective evaluation of the work performed in each individual task.

Questions

Below are questions associated with this chapter.

- Does your organization have an overall manpower planning model? If so, how is it implemented and used?
- Do you know what your team does and their task assignments?
- Do you have backup individuals assigned to the work done by your organization?
- Do you see that your organization may be restructuring or laying off individuals in the near future?
- Do you have another way that you gather and assess task assignments in your team that better fits your organization?

Reference

1. Tillman, Frank A., Cassone, Deandra T., *Manpower Requirements for Management and Professional Personnel*, Financial Times Press/Pearson Education, Upper Saddle River, NJ, Web. May 18, 2012

Aligning with Corporate Goals and Objectives

Abstract

You should develop a clear understanding of your organization's goals and objectives and how they are measured regardless of your position in the organization. You may not be part of an overall corporate strategic planning process, but information around where your organization wants to go strategically, and key measure of success should be available to you if you look for it. This chapter discusses concepts around establishing corporate objectives, developing measures to support those objectives, benchmarking, and presenting information to employees, peers and management. Aligning your work around known or derived corporate objectives ensures that you are focusing on the goals that are important to your organization. Knowledge, skills and workstyles identified in the first chapter that are ad-dressed in this chapter include the following:

Knowledge:

- Personnel and Human Resources
- Administrative
- Economics and Accounting
- Sales and Marketing

Expansion of core engineering characteristics at a broader organizational level:

- Judgment and Decision Making
- Critical Thinking
- Coordination
- System Evaluation

D. Cassone, *Leadership Concepts for the Engineering Mindset*, Synthesis Lectures on Engineering, Science, and Technology, https://doi.org/10.1007/978-3-031-67798-4_4

Skills:

- Management of Personnel Resources
- Social Perceptiveness
- Management of Financial Resources
- Negotiation

Understanding Corporate Objectives Supports Strategic Decision Making

A clear understanding of organizational objectives provides direction and focus for the corporation. Typically, resources are much more limited than the number of areas of opportunity for an organization to pursue, so you need to have a clear understanding of which opportunities best support the strategic direction of the organization. A clear picture and evaluation of the organization environment and the interaction between functions provides a framework to understand and evaluate the issues facing the corporation. Sound assessment of the issues within the company is critical to pinpoint key issues and opportunity areas. Measuring these key issues provides an understanding of the progress and success in improvement efforts.

You may not be positioned at an executive-level of the organization, but understanding what is important to your immediate organization and the organization provides perspective in your day-to-day operations. Understanding a broader view of your organizational objectives and current performance levels can be used to establish a vision for your own organization that supports the strategic direction of your company.

Recognizing and Defining Corporate Objectives

In a management or leadership position, should know where your company plans to go strategically. Executives may establish goals and objectives and communicate them to the organization, business unit and/or functional area and you be minimally involved in the process. Even though these goals and strategic objectives may be handed to you, you should use this information in your own area of responsibility to drive decisions that support corporate objectives. If your organization does not have clearly defined objectives, you can do your own research to better understand key goals and metrics for your industry, organization, or functional area. Aligning your decision process with organizational objectives helps support logical decision making within your domain of control.

Many strategic planning processes exist. The author and Dr. Frank A. Tillman present steps in their book, *A Professional's Guide to Problem Solving with Decision Science* [1, 2]. A clear understanding of organizational objectives and goals provides direction and focus for the corporation. Typically, there are many different areas of opportunity to pursue for improvement with limited resources. A clear picture and evaluation of the organization's environment and the interaction between functions provides a framework for understanding and evaluating the issues facing the corporation. A sound assessment of the issues within the company is critical to pinpointing the areas for improvement. Identifying and measuring these key areas provides an understanding of potential success in improvement. The process in the referenced book walks through the steps to identify corporate issues and develop solutions that can help the corporation improve. The structured thought process and solid evaluation of corporate functions and issues can help a corporation achieve improvement as measured by key performance metrics. This type of approach is important for developing meaningful objectives and developing means to measure success.

Even without a full strategic planning exercise, you can look at key measures that may be communicated to your organization by senior leadership. From the objectives, you can then develop meaningful measures for your organization that support the objectives. Examples of corporate objectives by different industries are shown below [1, 2].

- Service Companies
  - Improve Contract Performance
  - Minimize Operating Costs
- Manufacturing Companies
  - Minimize Manufacturing Costs
  - Maximize Customer Service
  - Minimize Distribution Costs
- Distribution Companies
  - Minimize Operating Costs
  - Reduce Inventory Levels
  - Improve Operating Cycle Time
- Insurance Companies
  - Maximize Customer Service
  - Streamline Information Technology
- Food Industry Companies
  - Improve Profitability
  - Increase Market Share
  - Increase Sales
  - Reduce Manufacturing Costs
  - Reduce Distribution Costs

These objectives are broad in nature and are used to focus management on organizational level goals. Organizational objectives provide the starting point for the decomposition of strategic goals into actionable activities and problem solving.

Organizational objectives provide the starting point for the decomposition of strategic goals into actionable activities and problem solving.

Depending on your industry and corporation, you may have publicly available data that can be gathered that addresses corporate objectives and the measurement of those objectives. For example, the following metrics are reported for an airline in the data that is provided to the Securities and Exchange Commission in their 10-k annual report. These metrics may be present on the company's financial statements, however, may not be known or understood at lower levels of the organization. Understanding these executive-level metrics will provide you with broader insights at your level of the organization. Key metrics across airline companies include the following.

- Size of fleet at year end.
- Number of Full Time Employees at year-end per Available Seat Mile.
- Number of Full Time Employees at year-end per Revenue Seat Mile.
- Number of destinations served.
- Salaries, wages and benefits per Available Seat Mile (ASM).

Spending time understanding the key objectives of your corporation provides you with a global view of operations that can then ensure your meaningful contribution to your organization. Having a deeper understanding of performance metrics reported by your company to various government and publicly entities help you to understand how your company may be performing in the industry overall and how your company's board of directors may view performance.

Engineers are drawn to metrics and quantifiable measures. You use quantifiable information when you are developing product specifications, constructing bridges, optimizing manufacturing operations, etc. Leverage your skills in quantifiable performance measurement at a managerial and executive level to provide you with measurable data from a management perspective. Measure and quantify those areas that most contribute to corporate success. Developing those measures based on corporate performance can be very valuable to your executive management team.

Metrics Management

Communicating your organizational performance through your team's metrics is necessary and critical to demonstrating your current operating level and performance. Most engineers love metrics, however, need to be thoughtful and strategic about the information that they collect and present to their management team. Your executive team should not need to fish through myriads of metrics to understand what is important to the organization. As an engineer communicating with leadership, you need to distill through the details to provide management with the most important metrics to them.

Metrics for Measuring Success

When an organization has defined and agreed upon its goals and objectives, decision makers can then develop a means to track and measure the accomplishment of these objectives. Often within an organization, a strategic plan, short-range plan, or long-range plan is developed and published as an edict to the organization. These plans may, however, end up on a shelf with no real meaning in the day-to-day operation of the organization. To drive these goals into the operating decision levels within the organization, these goals and objectives must be translated to meaningful measures of success that are used within the decision process and tracked within the organization. This is key to ensuring that the organizational goals are met.

Because so much data is currently available, you need to determine those measures (metrics) that provide a meaningful measure in achieving success within the organization. This process involves identifying decision criteria that are important to the organization. Typically, decisions are made based on a small number of criteria, so the metrics to support the decision process should be kept to a minimum.

Based on identifying the important criteria that support organizational objective, then data sources are assessed to determine whether this information can be relatively easily obtained, processed, and maintained. Data should be assessed to determine what level of detail should be maintained for it to be used as a metric to measure success. It may be tracked, for example, on a plant-by-plant basis or rolled up across the organization as a whole. Data can come from both automated sources or from assessments made by experienced executives or experts in the field. The process to establish, track, and maintain metrics and measures of success for an organization provides meaningful information in achieving success and understanding the organization's operations at every level. (Section reprinted from [1, 2].)

Understanding Metrics

A metric is a standard measure to assess performance in a particular activity. A metric is a composite of measures that yield systematic insight into the state of the process or products and drives appropriate action. Keep this in mind when you are developing and presenting metrics. Metrics can be composed of both objective data such as historical data reported in a database and subjective data such as expert opinions by senior management or experts in the business.

A metric is a composite of measures that yield systematic insight into the state of the process or products and drives appropriate action.

Metrics are important for several reasons. You can use them to defend and justify decisions, to provide objective assessment of progress toward goals, and for problem solving and to validate process improvement. Successful enterprises constantly assess themselves and improve in all dimensions of their enterprise. Metrics provide a foundation for the assessment of success of an activity or enterprise.

A good metric is built upon an organization's missions, goals, and objectives. A metric must be meaningful and understandable for management and workers alike. The best metrics derive naturally from the process in which the data is relevant to the process and its collection becomes part of the process. The metric itself must be easily measurable. The requirement for new systems and data to implement a metric should be minimal because if not, data will not be gathered to support it.

A company may choose any number of decision criteria and metrics to measure project success and evaluate the achievement of its goals. It is better, however, to measure against a few critical criteria rather than try to measure against interrelated criteria or less critical measures. Below are some metrics that you organization may use currently or could be considered for activities within your organization. Identify the most meaningful metrics for your organization.

Financial

- ROI (return on investment)
- ROA (return on assets)
- ROS (return on sales)
- Debt-to-equity ratio
- Inventory turns
- Units produced
- Units produced per hour
- Sales per employee
- Profit per unit
- Break-even volume

Operational

- Customer reported defects
- Production defects
- Repair defects
- Unscheduled maintenance hours
- Work in progress
- Number of machines per worker

- Changeover time
- Job classifications in the plant
- Obsolescence per model
- Energy consumption per unit
- Rate of absenteeism
- New product cycle time
- Engineering change requests
- Units batched
- Time metrics (response time, lead time, uptime, downtime, etc.)

Product Related

- Parts count
- Number of material types used
- Material use in each component
- Assembly process
- Service quality—repair/replacement
- Failure-mode effect analysis
- Customer quality
- Long-term durability of product
- Repeat customers
- Service response rates

Metrics are the recording and reporting of work done and outcome attained. Accomplished by collecting, storing, analyzing and systematically reporting information. Metrics measurement enables you to compare performance with standards. These comparisons are used to make adjustments to operations. Techniques such as time study, work sampling, and others can be used to collect data. (Section adapted from [1, 2].)

Benchmarking and Best Practices

Benchmarking and the use of key performance indicators, a balanced scorecard, or a dashboard are all similar. All these areas identify and quantify different performance characteristics of a company, typically within an industry. Benchmarking's primary focus is to compare internal performance with external benchmarked performance. Key performance indicators (KPIs) are the metrics deemed essential to understanding the operational health of a company. Measuring performance allows an organization to objectively determine what is working and what is not. The metrics or KPIs established for a benchmarking effort are compared to those of other companies to determine areas of potential improvement and to identify best business practices in an industry. Balanced scorecards and dashboards are the means by which KPIs are communicated to others.

Numerous companies provide intra-industry and cross-industry benchmark metrics. Developing and using benchmarks provides a concrete measure of your company's performance against others' performances. Books, industry organizations, companies performing benchmarking services, government data, and many other sources of information can be used to develop the hierarchy of criteria. The author has developed benchmarking metrics for the project management office at a telecom and has used industry standard metrics in a logistics management function. Understanding how your organization compares to other similar organizations provides value in day-to-day operations and is a great way to quantify your organization's performance across industry, identify areas for improvement and to provide a reference point to senior leadership on your organization's performance.

Those involved with benchmarking must avoid having a "one-size-fits-all" approach, which may not identify specific operating philosophies within a company. Not all types of "best practices" are "best practices" for all companies. You must be careful, though, to analyze the best practices of others in light of their culture and circumstances, or you may find that your efforts do more harm than good. Metrics should be tailored to capture data that is readily available across an industry.

Benchmarking identifies competitors that are doing better than your company. The goal then is to identify how other companies are achieving these better results. This leads to investigating best practices across industries to identify areas to improve operations and resulting performance. Benchmarking goes beyond the metrics and extends to the analysis, adoption, and development of best practices within your own company. (Section adapted from [3].)

Benchmarking goes beyond the metrics and extends to the analysis, adoption, and development of best practices within your own company.

Engineers can utilize benchmarking and best practice analyses to improve operations in their own organization. Leveraging engineers critical thinking and problem solving skills, you can provide best-in-class comparisons and the identification of better and best practices used by your competitors. Benchmarking exercises may uncover "low hanging fruit" for quick and impactful productivity improvements. From a leadership perspective, knowing how your competitors are performing and best practices they are using can demonstrate the breadth of knowledge that is necessary for executive leadership positions.

From a leadership perspective, knowing how your competitors are performing and best practices they are using can demonstrate the breadth of knowledge that is necessary for executive leadership positions.

Presenting Metrics

As you measure your organization, you will most likely need to present performance information to your employees, peers and management. Engineers are known to provide too much detail presenting results. The ability to present concise information to a group is an art and a science. You need to strike a balance between the information available and the information conveyed to the group. Engineers like to provide details as to how they arrived at their conclusions, but lengthy, non-focused presentations can be painful for the audience and lose key points and decisions required from the audience. You must spend time planning and practicing when you present information to others in your organization. Learning more about effective presentation skills and mastering those skills is important in your career development. Below are a few points to consider about presenting information.

- What is your purpose for the presentation?
- What do you want to accomplish?
- Do you know your audience and the level of detail that they expect to see?
- What is your audience's knowledge level?
- Do you have information and present it in such a way that you will reach your audience?
- Why are they attending?
- What do they expect from the presentation?

These considerations and how you organize and communicate your material on your presentation slides are important to reach your audience with the information and generate the decisions and results you intend from the presentation. Metrics should communicate performance and decisions, so should be viewed from the needs of the audience that you are addressing. Many tools are readily available, such as PowerBI and others that can assist in presenting your points. Unfortunately, good information can be lost in poor presentations.

Unfortunately, good information can be lost in poor presentations.

Questions

Below are questions for you to consider with regards to corporate objectives and metrics management.

- Do you know the goals and objectives of your organization?
- Have you reviewed any financial statements or industry analysis associated with your organization?

- Can you translate the corporate objectives to measurable metrics for your organization?
- Is benchmarking information available for your industry, company or functional organization?
- Do you have KPIs or dashboards developed for your organization's performance?
- Are the KPIs you track representative of your current environment, or should they be reevaluated?
- Do you have well developed presentation skills? Have you spent time planning your presentation to address your audience? Are there courses or resources that you can use to improve those skills?

References

1. Cassone, Deandra T., Tillman, Frank A., *A Professional's Guide to Problem Solving with Decision Science: An Integrated Approach for Assessing Issues, Finding Solutions, and Reaching Corporate Objectives*, Second Edition Overland Park, KS, Pioneering Partnerships, LLC, January 2018
2. Tillman, Frank A., Cassone, Deandra T., *A Professional's Guide to Decision Science and Problem Solving: An Integrated Approach for Assessing Issues, Finding Solutions, and Reaching Corporate Objectives*, Upper Saddle River, NJ, Financial Times Press/Pearson Education, Web: December 1, 2012, Hardcover: February 2012
3. Tillman, Frank A., Cassone, Deandra T., *The Science of Common Sense: Best Practical Decision Science Methods*, HTX, Inc., Osage Beach, MO, October 13, 2015

Budgeting for Your Organization

Abstract

Finance, accounting, and budgeting are not necessarily key focus areas of engineers and engineering core curriculums. Managing the financial resources under your control, however, is critical to your success. This chapter addresses the importance of managing your financial resources and introduces corporate budgeting and budget management. It is not intended to be comprehensive, but to introduce engineers to the budgeting process and key corporate financial statements that a company uses to manage its financial health. It is highly recommended that the engineer that is looking for more technical and management leadership positions enhance their knowledge in this area. Knowledge, skills and workstyles identified in the first chapter that are addressed in this chapter include the following:

Knowledge:

- Administrative
- Economics and Accounting

Expansion of core engineering characteristics at a broader organizational level:

- Judgment and Decision Making
- Coordination
- System Evaluation

Skills:

- Management of Personnel Resources
- Management of Financial Resources

D. Cassone, *Leadership Concepts for the Engineering Mindset*, Synthesis Lectures on Engineering, Science, and Technology, https://doi.org/10.1007/978-3-031-67798-4_5

Importance of Managing Financial Resources

One of the quickest ways to demonstrate your organizational management and leadership skills is to effectively manage your budget. Financial management is a quantifiable way to show that you are managing your resources. Good financial management of your immediate area of responsibility can position you for broader levels of financial responsibilities. There are two examples that the author has seen that have demonstrated this importance, one of which was in industry, the other was in an academic position.

One of the quickest ways to demonstrate your organizational management and leadership skills is to effectively manage your budget.

While working as an operations manager in a Fortune 100 company reporting to a new Vice President of Reverse Logistics, this executive came into the company with a laser focus on finance. This individual had been brought in by the new corporate Chief Executive Officer to drive operational changes in the organization. Previously, he had been the CEO of a company. His finance team was of utmost importance to him, and the direct reports spent hours developing enhanced financial reporting and reviewing all the expenditures of his organization. The vice president wanted to be sure that his financial house was in order and expended the resources to demonstrate that it was in order. From this focus, it became evident that there was a significant amount of repair part inventory. This inventory investment was in the millions of dollars and had not been actively managed at this point. The vice president reviewed the parts in inventory and then began selling inventory that was unnecessary to support the current repair requirements. He reduced inventory levels by millions and enabled the company to expend those resources elsewhere.

The second example is from an academic perspective. The author was director of operations in a research center. If you've not been exposed to the academic world, you might think that university leaders may not be focused on finances. This is anything but the case. When the academic leader moved to a new university and organization, a key focus of the leader was to understand his budget and the resources available to support his research center activities. This involved understanding not only the budget amounts to support the new position, but understanding the personnel and operating allocations for each of the grants and contracts that supported the center. This leader wanted to know the current allocation of effort on grants, his startup funding and operating costs so that he would know his financial position and plan for future support of the center. Additionally, he quickly established a service center that could bill other internal and external entities on a fee-for-service basis so that as other entities wanted to leverage the center resources, there was a way to bill for the use of these resources for those billable services. It was very interesting to see this laser focus on budget. By spending a significant amount of time establishing budgets, allocating personnel effort, assessing current grant funding, projecting future grant funding based on proposals submitted, he could see his current financial position and project his future growth opportunities. This enabled him to expend resources and implement state-of-the-art technology to support initiatives on a local and national scale.

Corporate Budgeting and Cost Centers

A budget is a formal written statement of management's plans for a specified future time period, expressed in financial terms. A budget is a primary way to communicate the agreed upon objectives of an organization to executive leadership, management, peers and employees. Understanding budgets promotes efficiencies by having clearly defined financial objectives established. It is also a great control device and is an important basis for performance evaluations once adopted.

Budgeting has many important benefits as listed below.

- Requires all levels of management to plan and formalize goals on a recurring basis.
- Provides definite objectives for evaluating performance at each level of responsibility.
- Creates an early warning system for potential problems.
- Facilitates coordination of activities within the business.
- Provides for future planning for an organization.
- Allocates resources across divisions and units in an organization.
- Provides for the coordination of activities.
- Communicates plans in quantifiable terms.
- Results in greater management awareness of overall operations and external factors.
- Motivates personnel to meet planned objectives.

Accounting obviously has a key responsibility in developing budgets, however, it is not the accounting department's responsibility to manage budgets. Historical accounting data helps in formulating future budgets. Accountants are typically responsible for presenting management's budget in financial terms. The budget and its administration are, however, entirely management's responsibility.

Accounting has a key responsibility in developing budgets, however, it is not the accounting departments responsibility to manage budgets. The budget and administration are management's responsibility.

Developing Budgets

Figure 1 shows the flow of budgeting information. Many times, management believes that budgeting information flows only from top management to lower levels of management. They may not understand that they have influence in the budgeting process. Top

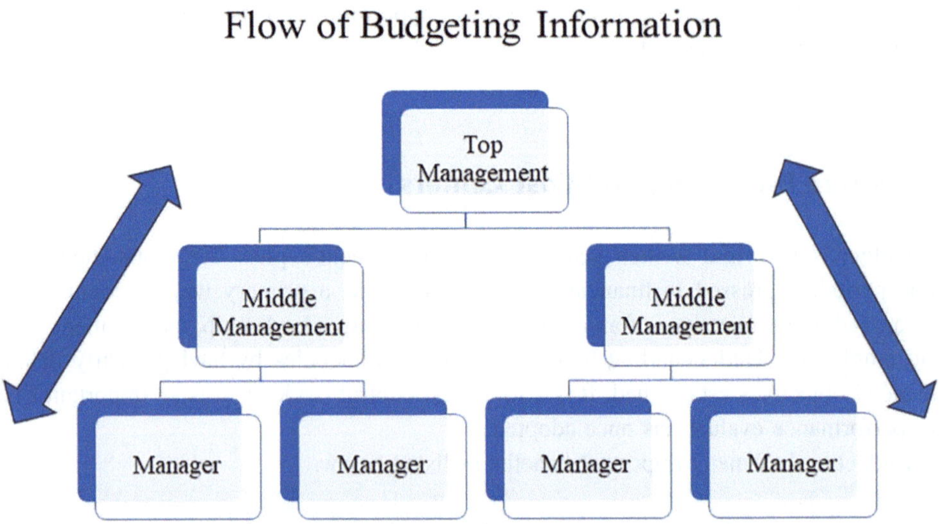

Fig. 1 Flow of budgeting information

level management uses budget information from lower levels of management to formulate their budgets and projections for the following year. Operational expenditure will drive overall organizational resource expenditures so the costs and resources used at an operational level will drive allocations for future budgets. That's why understanding and managing your budget, personnel expenditures and operational expenditures can impact the budgets you receive in the following years. You can also provide valuable information in areas of operational efficiencies and project expenditures that can improve your domain's utilization of resources.

Operational managers and executives should not only manage their budgets but optimize their budgets and identify areas to improve financial performance. You can assist your organization's holistic financial performance by identifying areas where costs can be reduced and developing business cases for projects that can improve overall operations. What may seem like pie in the sky numbers that you receive for your organizational budget can be impacted by your communication with management on key expenditure and operational improvement areas [1, 2].

Operational managers and executives should not only manage their budgets but optimize their budget and identify areas to improve financial performance.

Figure 2 shows an example of the types of information and budget development process for a manufacturing company. The typical budgeting process is very detailed and contains quantifiable financial information about each of the components of the budget. This section is not presented to provide a complete view of financial management in an

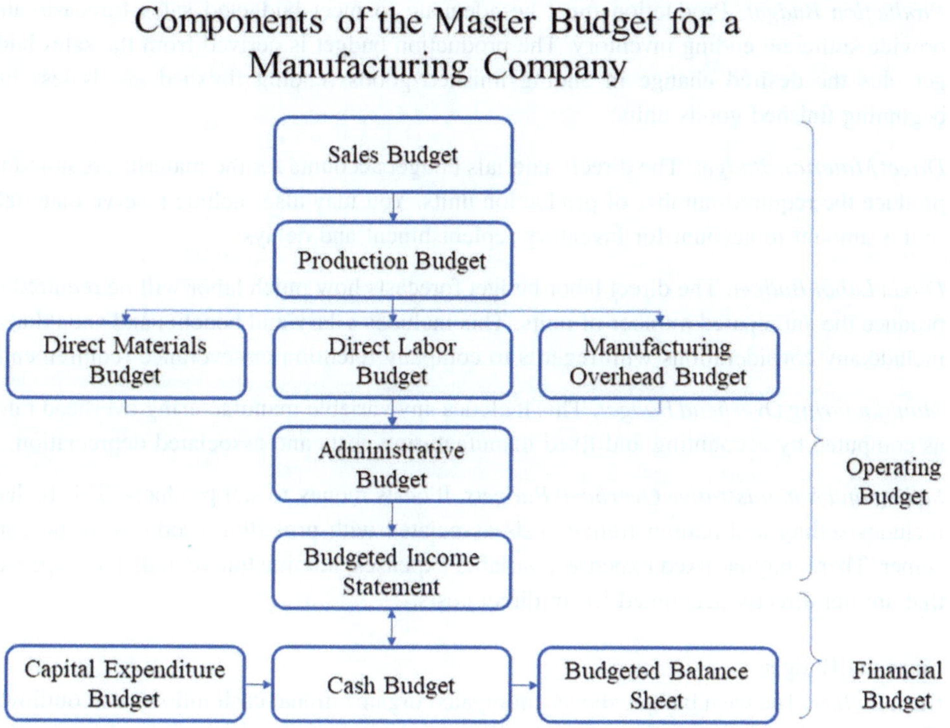

Fig. 2 Example of the budgeting process

organization, however, is provided to understand how from various organizational levels the operating processes and costs impact the overall financial position of a company. More knowledge can be gained from additional study in finance and accounting [1, 2].

Operating Budget

Sales Budget. Sales budgets drive the remainder of the budgeting process. The sales budget is a detailed schedule showing expected sales for the budgeted periods expressed in units and dollars. The sales budget is prepared first and derived from the sales forecast. Management develops their best estimate of sales revenue for the budget period. Every other budget depends on the sales budget. The sales budget is prepared by multiplying the expected unit sales volume for each product times anticipated unit selling price. Factors considered in creating a sales forecast include general economic conditions, industry trends, previous market share, price changes, technological developments and other sales-related considerations.

Production Budget. Production must be adequate to meet budgeted sales forecasts and provide sufficient ending inventory. The production budget is derived from the sales budget plus the desired change in ending finished goods (ending finished goods less the beginning finished goods units).

Direct Materials Budget. The direct materials budget accounts for the materials required to produce the required number of production units. You may also include reserve materials in this amount to account for inventory replenishment and delays.

Direct Labor Budget. The direct labor budget forecasts how much labor will be required to produce the anticipated number of units. This includes salary and benefits and should also include any considerations with regards to company retention or severance requirements.

Manufacturing Overhead Budget. This includes any variable manufacturing overhead rates as computed by accounting and fixed manufacturing costs and associated depreciation.

Selling and Administrative Overhead Budgets. It costs money to sell products. This budget includes selling and administrative costs associated with providing products to the consumer. There may be fixed expenses, variable expenses, administrative and staff expenses that are not directly accounted for in direct costs.

Financial Budget

Cash Budget. The cash budget shows anticipated organizational cash inflows and outflows. The cash budget is very often considered to be the most important output in preparing financial budgets. Cash keeps a business running and without it, or sources of financing, a company cannot survive. The cash budget consists of cash receipts, cash disbursements and financing. The cash budget contributes to more effective cash management and shows managers the need for additional financing before the actual need arises. Additionally, the cash budget indicates when excess cash will be available. Beginning and ending cash balances are shown on financial statements.

The cash budget is very often considered to be the most important output in preparing financial budgets. Cash keeps a business running and without it, or sources of financing, a company cannot survive.

Budgeted Income Statement. From the development of the cash budget, the budgeted income statement can be derived. The budgeted income statement shows the anticipated results of a company's operations over a time period. This statement shows what goods were sold or services performed that provided revenue for the company. It also shows what costs were incurred in normal operations to generate these revenues. Overall, it shows the earnings and profit of a company. Key components of the budgeted income statement are company revenues, company expenses and the net profit of a company. Engineers that understand the operating components associated with the income statement can assess their operating domain and see how their work functions impact the company's income statement.

Engineers that understand the operating components associated with the income statement can assess their operating domain and see how their work functions impact the company's income statement.

Budgeted Balance Sheet. The budgeted balance sheet is classified and comparative. There is more than one balance sheet compared so that executives can identify significant changes over time. The balance sheet distinguishes between current and long-term assets, current and long-term liabilities listed in decreasing order of liquidity. The budgeted balance sheet highlights what the organization owns, what they owe to creditors and the owners claim against resources as shown in the Accounting Equation: Assets = Liabilities + Owners' Equity. Limitations of the balance sheet include the fact that assets are recorded at historical value, it only recognizes assets that can be expressed in monetary terms and owners' equity is usually less than the company's market value. This is why it is necessary to assess the income statement, balance sheet, cash flows and other financial reports together to gain a full picture of company health.

The budgeted balance sheet highlights what the organization owns, what they owe to creditors and the owners claim against resources as shown in the Accounting Equation: Assets = Liabilities + Owners' Equity.

Managing Budgets

"One of the most important finance skills for managers to master is budgeting, or the process of preparing and overseeing a financial plan that estimates income and expenses over a defined period." Harvard Business School Online's blog identifies six important budgeting tips for managers which have been summarized below [3].

1. **Know Your Organization's Budgeting Timeline and Procedures**. Each organization is different, and you should understand the budgeting process and timeline for your organization. Some organizations use a yearly budget, some have an eighteen-month budget, and some organizations use a rolling budget that is perpetually updated. Understand what budget items are tracked for your organization and measure those items on an ongoing basis. Financial and operating numbers can be tracked as part of your key performance indicators.
2. **Leverage Financial Data**. Understand the key financial statements for your organization. This may include your company's reported balance sheets, income statement, statement of cash flows, retained earnings, etc. You can look up your organization's financial statements, if your company is publicly held, on the Security and Exchange

Commission's website. Many times, organizations will publish their financial statements on their website as well. When reviewing financial statements, read the narrative associated with the statements. This provides context and explanations associated with the statements and familiarizes you with company direction that may be communicated to stakeholders.

3. **Work Toward Goals**. As discussed in the previous section, understanding the specifics around operating costs for your organization and how your organization builds its budget can assist you in pinpointing specific areas that you can impact. You can use these operating and financial characteristics to assign resources and team members to improving operations. You can also communicate specific improvements to management and how they impact the corporate budget as well as provide specific areas of performance or necessary improvement.
4. **Evaluate Performance**. Measuring specific costs and performance that contributes to these costs can pinpoint areas for employee evaluation and effectiveness. You need to understand the financial implications of your operating performance. You can then translate this to employee accountability and performance. If you don't have specific operating and financial measures for performance evaluation, you may need to lean on more subjective evaluations instead of quantifiable performance measures.
5. **Communicate Progress and Results**. Develop a cadence for updating your budget and performance numbers and communicate those results to your employees and management chain. Most budget and performance-related numbers are available monthly. You should update your budget to actual performance numbers when they become available so that you can then see issues early and reward performance improvements. You should also determine when it is appropriate to communicate your performance numbers to others in your organization. You may want to schedule quarterly performance reviews with broader stakeholders. This will allow you to recognize changes in performance and provide the performance numbers with operational changes that you may have made. Determine your reporting cadence based on what is best for your organization and reporting relationships.
6. **Hone Your Financial Knowledge and Skills**. Learn more about organizational financial reporting and gain additional knowledge in corporate finance. This knowledge will assist you in your day-to-day operations and assist you in making informed business decisions. Demonstrating financial acumen will also support your movement to positions with leadership responsibilities.

Budgeting and financial management skills may not be readily taught in engineering curriculums but is a natural extension of your operational responsibilities. Understanding the financial representation of your work provides a "common" communication language with others outside of your technical discipline. Ultimately, companies think in financial terms and translating your operational responsibilities into this common language will help you communicate your performance to the organization at large.

Budgeting and financial management skills may not be readily taught in engineering curriculums but is a natural extension of your operational responsibilities.

Capital Budgeting and Business Case Development

When engineers identify opportunities for improvement, replacement, or innovation that require large financial outlay, they enter the area of capital budgeting. Typically, this involves the analysis of potential projects or long-term decisions that will impact their firm's resources. Expenditures in these areas can be large, so specific analysis is needed to justify the expenditure of resources. Corporate management will need to understand the revenue that the organization will gain, the expenses that they will incur and the financial and strategic benefit that they will receive from capital expenditures.

Corporate management needs to understand the revenue that the organization will gain, the expenses that they will incur and the financial and strategic benefit that they will receive from capital expenditures.

Business cases are developed to justify these expenditures. Some engineering disciplines require a course on engineering economy that describes the computational procedures for estimating the return on your resource investment. Basic computations require you to estimate the cash inflows and cash outflows over a given time period at a given cost of capital or interest rate to determine if the expenditure will pay back the investment. Methods of estimating the expenditure value are measured in terms of its payback period, net present value and internal rate of return that represent cash inflows and outflows in a common measure of comparison. Strategic considerations along with financial considerations can influence the decision to pursue a capital project. Engineers should familiarize themselves with these computational processes so that they can calculate and/or provide input into business cases that justify capital expenditures.

Questions

Below are questions to consider from this chapter.

- Do you have direct budget management responsibilities, or do you support your managements budget management responsibilities?
- Do you understand the budgeting process used in your organization?
- Can you identify areas in your domain of control that you impact directly?
- Do you have knowledge of financial statements such as income statements, balance sheets, statement of cash flows and others?
- Have you reviewed your organization's financial statements?

References

1. Garrison–Noreen–Brewer: Managerial Accounting, 11th Edition, The McGraw–Hill Companies, 2006
2. Engineering Management, Meeting the Global Challenges, Second Edition, by C.M. Chang, CRC Press, 2016
3. Gavin, M., 6 Budgeting Tips for Managers, April 23, 2020, HBR Blog 6 Budgeting Tips for Managers | HBS Online

System Evaluation of the Organization

Abstract

System evaluation and operational analysis is a knowledge area and skill that has been identified across many engineering disciplines. Engineers design components and evaluate systems based on their own technical disciplines, however, those designs and components must interface with the system as a whole. Systems engineering process can translate to the evaluation of an organization, overall. Systems engineering concepts can be translated into enterprise systems engineering and provide a useful view of your organization and its interfaces with functions within an organization. Viewing your organization from a system's perspective and understanding the interrelationships and impacts between functions enables you to take a holistic approach to management and problem-solving. To address organizational functions and interfaces can require a significant amount of work as well as the commitment of senior leadership. The results, however, are a more holistically optimal approach to problem solving that can provide benefits to the organization as a whole. Knowledge, skills and workstyles identified in the first chapter that are addressed in this chapter include the following:

Knowledge:

- Personnel and Human Resources
- Administrative
- Economics and Accounting
- Sales and Marketing

Expansion of core engineering characteristics at a broader organizational level:

- Judgment and Decision Making
- Complex Problem Solving

D. Cassone, *Leadership Concepts for the Engineering Mindset*, Synthesis Lectures on Engineering, Science, and Technology, https://doi.org/10.1007/978-3-031-67798-4_6

- Critical Thinking
- Coordination
- System Evaluation

Skills:

- Management of Personnel Resources
- Management of Financial Resources

Systems Engineering

The International Council on Systems Engineering (INCOSE) is a professional society that supports and promotes the field of systems engineering. INCOSE produces and maintains the online Systems Engineering Book of Knowledge (SEBoK) which is a wiki-style reference open to contributions but maintained by an editorial board. Systems engineering is a logical, structured approach to managing complex systems. The concept of systems engineering is a useful overlay of many technical engineering disciplines and provides a framework for logically interfacing and integrating complex systems into an operationally whole system. Many engineers think in systems.

Exposure to systems engineering is useful on many levels. From a complex project management perspective, managing complex systems projects takes a significant amount of rigor to ensure the complex interfaces work together. Systems engineering also addresses holistic life-cycle costs of systems such as design and development, operations, maintenance, retirement and disposal, not just initial investment costs that are typically used in business case development and justification. Systems engineering can also provide a context for managing the complex "systems" of an organization with enterprise systems engineering. There may be internal complexity in the functional organization but organizations should work to maintain simpler interfaces with other functional organizations. Enterprise systems engineering provides a useful view of your organization and its operations and interfaces. As shown in the first chapter, engineers have skills in Operations Analysis, Systems Analysis and System Evaluation. CEO's skills also include System Evaluation. Engineer's skills in system evaluation at a lower level of the organization can be expanded and leveraged in management and executive positions.

Engineer's skills in system evaluation at a lower level of the organization can be expanded and leveraged in management and executive positions.

INCOSE provides the following systems engineering related definitions [1].

A system is an integrated composite of people, product, and processes that provide a capability to satisfy a stated need or objective. Systems engineering is an interdisciplinary engineering management process that evolves and verifies an integrated, life-cycle balanced set of system solutions that satisfy customer needs. The three essential definitions of Systems Engineering published by the International Council on Systems Engineering (INCOSE) are shown below.

Systems Engineering. Systems Engineering is a transdisciplinary and integrative approach to enable the successful realization, use, and retirement of engineered systems, using systems principles and concepts, and scientific, technological, and management methods. We use the terms "engineering" and "engineered" in their widest sense: "the action of working artfully to bring something about".

Engineered Systems. "Engineered systems" may be composed of any or all of people, products, services, information, processes, and natural elements. An engineered system is a system designed or adapted to interact with an anticipated operational environment to achieve one or more intended purposes while complying with applicable constraints. Thus, an "engineered system" is a system – not necessarily a technological one - which has been or will be "systems engineered" for a purpose.

System. Finally, a very general and inclusive definition of "system" is provided: A system is an arrangement of parts or elements that together exhibit behavior or meaning that the individual constituents do not. Systems can be either physical or conceptual, or a combination of both. Systems in the physical universe are composed of matter and energy, may embody information encoded in matter-energy carriers, and exhibit observable behavior. Conceptual systems are abstract systems of pure information, and do not directly exhibit behavior, but exhibit "meaning". In both cases, the system's properties (as a whole) result, or emerge from: the parts or elements and their individual properties; AND the relationships and interactions between and among the parts, the system and its environment [1].

Below is a brief definition of the key steps in the System Engineering process with a summary view in Fig. 1 [2] which can be both sequential and iterative. The systems engineering process assists in decomposing complex problems, developing solutions, iterating through when introducing new requirements, and validating the successful operation of a system. The steps listed below provide understanding in the components of the systems engineering process. Again, these concepts are only provided as an introduction to this field of study and application. These concepts can also be applied to organizational operation with Enterprise Systems Engineering which is discussed in the following section.

- **Definition of Problem**—Define the nature and magnitude of the problem, risks and express them in system requirements.
- **System Feasibility Analysis**—Evaluate different technological approaches to meet the specified functional requirements.

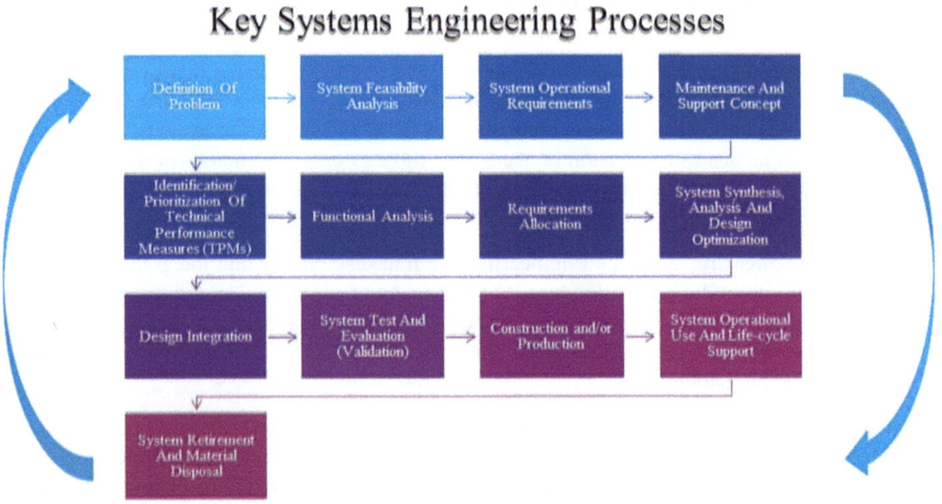

Fig. 1 Summary of key systems engineering processes

- **System Operational Requirements**—Identification and synthesis of the functions required, performance, quality measures, assessment criteria and verification of the completed system.
- **Maintenance and Support Concept**—Defines the levels of maintenance required, repair policies, Organizational responsibilities, support, effectiveness requirements and environment.
- **Identification/Prioritization of Technical Performance Measures**—Review of requirements in terms of specific quantitative values, their degrees of importance and criticality in accomplishing the mission of the system. Cost and system effectiveness should be included in the analysis.
- **Functional Analysis**—Identify the resources necessary for the system to accomplish its objectives and break down requirements from the system level to the subsystem, to the degree necessary in the hierarchical structure.
- **Requirements Allocation**—Define the specific input design criteria for the various subsystems and major lower-level components of the system and allocate or apportion to the major sub-components.
- **System Synthesis**, Analysis and Design Optimization—Combine and structure components for a feasible system configuration leading to the definition of possible alternatives that can analyzed, assessed and evaluated to determine the best system architecture.
- **Design Integration**—Provides a "big picture" view of the system created with integrated and cross functional design teams.

- **System Test and Evaluation (Validation)**—Comprehensive planning required for successful testing effectiveness with multiple categories of test and evaluation based on the stage and progress of the system development life cycle.
- **Construction and/or Production**—System development including production and construction with inherent design elements that account for the constructability and producibility of the system.
- **System Operational Use and Life-Cycle Support**—Design for the entire system life cycle including sustaining maintenance and support and assessment of new technologies and modifications for improvement. Data is used to monitor and improve the system operational capabilities on an ongoing basis.
- **System Retirement and Material Disposal**—Address the entire system life cycle accounting for retirement and disposal requirements including the potential repurposing and reutilization of components and parts.

Enterprise Systems Engineering

The engineering manager or executive can utilize systems engineering concepts when evaluating how the organization operates overall and how their functional organization operates within the context of the corporate system overall. The below definition and description provide context on the enterprise systems engineering concepts.

> A relatively new discipline called Enterprise systems engineering (ESE) has evolved and is being used in enterprises to enhance organizational processes.
>
> Enterprise systems engineering (ESE) is the discipline that applies systems engineering to the design of an enterprise. As a discipline, it includes a body of knowledge, principles, and processes tailored to the design of enterprise systems. An enterprise is a complex, socio-technical system that comprises interdependent resources of people, information, and technology that must interact to fulfill a common mission. Enterprise systems engineering incorporates all the tasks of traditional systems engineering but is further informed by an expansive view of the political, operational, economic, and technological (POET) contexts in which the system(s) under consideration are developed, acquired, modified, maintained, or disposed] An enterprise systems engineer must produce a different kind of analysis on the people, technology, and other components of the organization in order to see the whole enterprise. As the enterprise becomes more complex, with more parameters and people involved, it is important to integrate the system as much as possible to enable the organization to achieve a higher standard.
>
> Figure 2 shows how these business activities relate to each other as well as the relative scope of ESE and product systems engineering (PSE). PSE is mainly involved at the project level and collaborates with project management activities and gets somewhat involved in program management and the resource allocation and budgeting activities. On the other hand, ESE is heavily involved in the higher-level activities from the program management level and up. Both ESE and PSE have key roles to play in the enterprise [3].

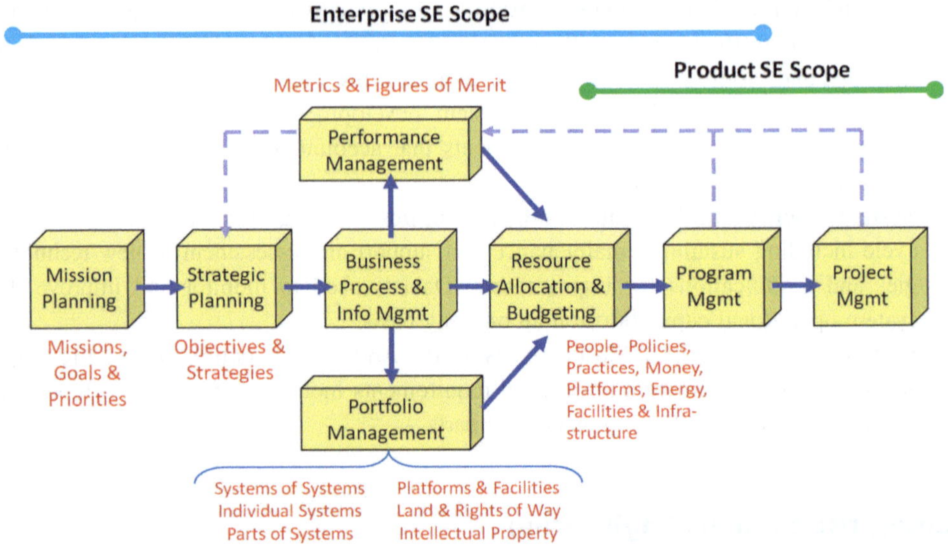

Fig. 2 Enterprise and product systems engineering scope (https://sebokwiki.org/d/images/sebokwiki-farm!d/0/06/ESE-F09.png)

> Shown in this manner, these business activities can be considered to be separate processes with a clear precedence in terms of which process drives other processes. Technical Systems Engineering (TSE) uses "requirements" to specify the essential features and functions of a system. An enterprise, on the other hand, typically uses goals and objectives to specify the fundamental characteristics of desired enterprise operational capabilities. The enterprise objectives and strategies are used in portfolio management to discriminate between options and to select the appropriate balanced portfolio of systems and other enterprise resources [4].

Systems engineering and its knowledge area of Enterprise systems engineering are fields of study. These processes and applications will not be covered in detail in this book but are provided for suggesting an approach on how to look at your organization. What are the objectives? What are the interfaces with other organizations? How do employees, resources and business processes interact within and between organizations? The concepts and processes used in systems engineering and enterprise systems engineering can be useful when evaluating and assessing your organization.

> **The concepts and processes used in systems engineering and enterprise systems engineering can be useful when evaluating and assessing your organization.**

Engineering-Centric View of the Organization

Definitions and discussions about key functional organizations were discussed in the first chapter and shown in Fig. 8 in that chapter. Figure 3 highlights typical operational areas that an engineer may support in the context of overall corporate operations. These areas typically include responsibilities in design and development, production and construction, warehousing and logistics, distribution and supporting the information technology infrastructure of an organization.

Each organization is different in how they organize themselves and operate within the various departments and divisions. Engineers understand and know that there are many corporate functions outside of their engineering operating domain. Understanding how you interface with corporate functions outside of the typical engineering functions is important to have a better understanding of how you best can leverage and interact with these functions. Understanding how your organization interacts and interfaces with other functions of the organization can be critical to your success. Important here is seeing your function and organization in the context of the overall organization. It takes all these functions to run a company and as you move into management and leadership positions, it becomes more important to recognize these functions and understand the importance of interacting with them.

Understanding how your organization interacts and interfaces with other functions of the organization can be critical to your success.

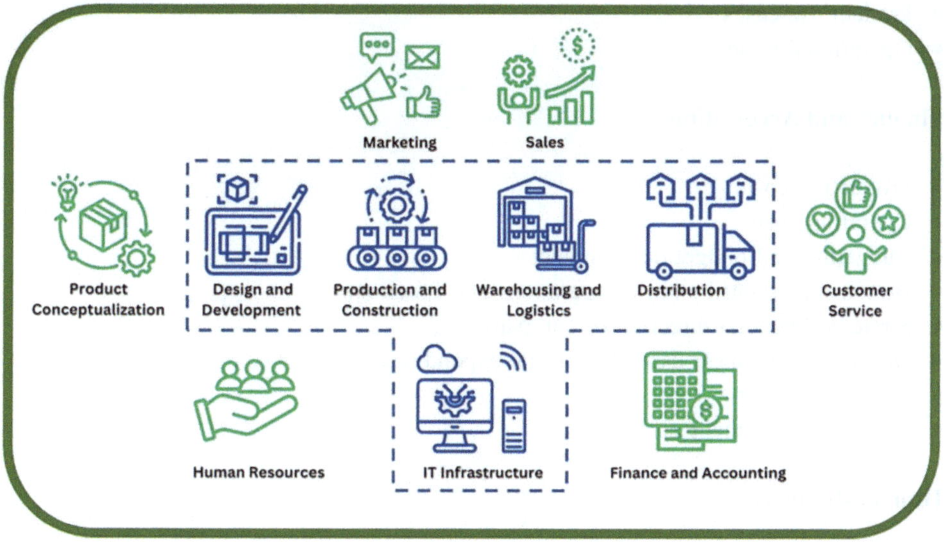

Fig. 3 Typical engineering functions with expanded view

Below are some ways that engineering functions may interact with other functional areas of an organization.

Marketing

- Designing products that meet customer needs.
- Communicating product features
- Communicating product limitations
- Developing marketing materials

Sales

- Product operation
- Customer feedback on products
- Developing customer service materials
- Developing users' manuals
- Facilitating implementation and training
- Operating products
- New feature suggestions

Customer Service

- Detailed product operations
- Customer feedback
- Product use cases
- Operational issues

Finance and Accounting

- Production costs
- Operating costs
- Budgeting and budget development
- Bottom-line contribution of operations and functions
- Strategic financial direction of company
- Developing business cases for capital expenditures
- Vendor selection
- Vendor management

Human Resources

- Development of "soft skills"
- Development of leadership skills

- Employee performance evaluations
- Employee actions such as promotions, hiring, firing and layoffs.
- Employee development
- Legal requirements for employee management
- Leadership training
- Employee rules, processes and procedures

Product Conceptualization

- Opportunities to support new product development.
- Working with cross-functional teams to develop product roadmaps.
- Working with market researchers to identify new opportunities.
- Developing specifications for new products
- Integrating marketing and sales feedback into product design, development, implementation and delivery
- Innovation

Engineering functions are a part of overall organizational operations. As you move into greater levels of technical and managerial leadership positions, you will be interfacing with the broader organization. Understanding how to best interface and building broader functional relationships will benefit you in your career.

Building broader functional relationships will benefit you in your career.

Clustered Evaluation of Organizational Issues and Opportunities

Many different corporate planning methods focus on identifying strategic objectives and then developing plans to accomplish those goals. These methods help senior management to determine focus and direction for future activities within a company. This section describes a method that focuses on identifying opportunity areas based on the interrelationship of functions within an organization. The intent of this section is to provide you with one approach that you can use to develop cross-functional solutions.

Corporate planning and enhancements can be done from a stove piped perspective, but many organizational issues exist across functional disciplines. This approach focuses on problem areas and interrelationships that need to be addressed to determine areas of the company that need to be fixed to move it into a best-in-class status. From a problem-focused corporate planning perspective, this process consists of the assessment of the relationships between the key areas of a company, an assessment of each of those key areas in terms of their performance against industry benchmarks, and then visually represents these relationships and opportunity areas that need to be included in the analysis to improve company performance. The book, "*A Professional's Guide to Problem Solving Using Decision Science*", Cassone and Tillman, contains a full description and examples in the use of this method [5, 6]. A summary of analysis steps that can be used to perform this evaluation are provided below.

1. **Assess the Scope of the Problem**—In this step, key corporate functions are identified for evaluation. Activities in one functional area may or may not have an impact on other functional areas. Certain functional areas impact all areas of the company such the Enterprise Resource Planning system or IT infrastructure of an organization.
2. **Represent the Relationships Between Functions with an Activity Relationship Matrix**—This step captures the interactions between the functions. An assessment is made to qualify those interactions. This assessment can be recorded in an activity relationship matrix which shows the strength of the association between functions. Developing a diagram to provide a visual representation of the strength of interaction between related areas can also be helpful.
3. **Quantify Performance with Industry Benchmarks and Performance Evaluations**—In this step, key performance indicators or decision criteria derived from corporate objectives are used to evaluate the performance of each of the functional areas. This provides an assessment of relative performance with industry. These metrics are established standards to quantify an overall representation of the problems and their benchmark performance with each of the functional areas.
4. **Develop the Activity Relationship Diagram**—This diagram provides a view of the interaction between functions and industry benchmarks and indicates the opportunities for improvement in each of the areas. A relationship convention is used for the diagram, and which functional areas represent the greatest opportunities for improvement.
5. **Determine the Variability of the Process and Financial Contribution of the Individual Functions**—The variability analysis measures the level of control and the performance of key functional areas. The percent of the total corporate budget for each of the functional areas is also assigned to indicate the financial value of improvements of this function. This provides a measure of the performance improvements and its impact to the corporate bottom line and the operational and financial performance.
6. **Identify Specific Problem Areas to Improve**—This analysis provides a roadmap to identify specific problem areas in the company to improve from a cross-functional perspective. This leads to the scope of the analysis and the individual decision criteria and metrics that show the focus areas for issue resolution.

These analysis activities are directed toward an executive-level evaluation of the company and its functions. Resources must be managed to provide the best return on human and monetary capital. The approach is intended to develop a high-level integrated view of where to expend organizational resources with the greatest impact, thus avoiding developing stove-piped solutions. This holistic view can then be used to drill down into specific areas in which problems need to be solved. The integrated view is provided because it is important to provide a representation of the interactions between organizational functions and it also quantifies some indication of magnitude of the issues a company may face This approach provides a more system-driven and functionally integrated approach to understanding and improving company operations.

A high-level integrated view of where to expend organizational resources with the greatest impact will avoiding developing stove-piped solutions.

Building a Panoramic View of the Process

From a systems evaluation perspective, you need to be able to communicate succinctly the issues that you are looking to resolve. How can you get a panoramic view of the process? With the building blocks of objectives, metrics, functional interactions, and functional performance, you still must pull together all the pieces into one view of the process. Business process modeling provides a tool to understanding the holistic picture. No activity stands alone in nature, and the inputs, outputs, and interactions between the activity are critical to document. Create a one-page view of the activities, influences, systems, and results that surround this process. This type of process incorporates the following key elements [4, 5].

- Identification of the problem area and the scope of the problem.
- Definition of the sphere of control.
- Identification of the key activities, processes, systems, and influences of the problem area.
- Upstream and downstream interactions between the areas and activities.
- Identification of the data that supports the measurement of the metrics.

A high-level business process model provides a roadmap for analyzing the company activities. Sometimes you need to delve into the subcomponents or the sub subcomponents of the problem to discover the issues and answers required to solve the problem. Detail designs associated with components and computer programs do require specific, detailed documentation; however, generally, high-level business processes can be represented with a 20-to-30 block diagram that provides a clear overview.

Upper management attitudes can drive operational behaviour in identifying certain problems and developing potential solutions. You must be aware that treading on analysis and solutions that may put upper management in a less than favourable light to the organization can be impactful to a career. Because of this, you must understand the political organizational environment enough to know where and who the potential solutions may impact. Based on your political organizational environment, you must also know how to propose the solutions in a way that has the best chance of being implemented. This knowledge must guide the problem solver to gain acceptance of the actual solutions by management [4, 5].

Based on your political organizational environment, you must also know how to propose the solutions in a way that has the best chance of being implemented.

To develop solutions that provide integrated value to the organization, you will also need to engage other leaders within the organization to get their buy-off on proposed solutions. In your current position, you may only have direct influence on limited number of processes and individuals that report directly to you. To develop implement solutions that impact your functional area but are not in your direct control requires that you understand those functions impacting you. Understanding the sphere of control in your functional area and those areas that you have close functional relationships will assist in developing cross-functional solutions [4, 5].

Additionally, you must understand the financial climate of an organization. Are funds available for major or even minor changes? If so, projects can be initiated for system enhancements, construction projects, or new marketing campaigns. If funds are not available, implementable solutions may need to be tempered so that they are low- or no-cost solutions. This adds an additional layer of complexity to the problem solver and the solutions generated.

Fundamentally, you can view your organization as an enterprise system with functional complexity and interfaces with other organizations. Viewing your organization from a system's perspective and understanding the interrelationships and impacts between functions enables you to take a holistic approach to management and problem-solving. This chapter is only intended to introduce these topics. To address organizational functions and interfaces can require a significant amount of work as well as the commitment of senior leadership. The results, however, are a more holistically optimal approach to problem solving that can provide benefits to the organization.

Questions

Below are questions to ask yourself regarding this chapter.

- Is your organization familiar with systems engineering techniques? Do they use these techniques in project development?
- Can you think of your organization in the context of enterprise systems engineering? Can you view the functional areas in your corporation as a system of systems?

- Develop a view of your holistic organization from a system of systems perspective?
- What are other ways that organizations outside of your engineering functions impact engineering functions? The list above is not comprehensive and there may be other interactions in your organization not included in the list.
- Have you evaluated cross-functional or clustered solutions to issues facing your organization? Do you have good relationships with individuals from other functional areas?
- Is your organizational environment open to change or protective of functional areas?
- Have you created high-level business process models for your organization or have they been done previously and available for review?
- What is your sphere of control? What is your boss's sphere of control?

References

1. Systems Engineering and System Definitions, Version: 1.0, Issued on: 8 January 2019, AUTHOR TEAM, Hillary Sillitto, James Martin, Dorothy McKinney, Regina Griego, Dov Dori, Daniel Krob, Patrick Godfrey, Eileen Arnold, Scott Jackson, Copyright (c) 2019 by INCOSE, published under Open Access. https//:System and SE Definitions (incose.org) Reprinted with permissions
2. Blanchard, B., Blyer, J., System Engineering Management, Fifth Edition, John Wiley and Sons, 2016, Hoboken, New Jersey
3. Guide to the Systems Engineering Body of Knowledge (SEBoK), version 2.9 Released 20 November 2023 Please note that this is a PDF extraction of the content from www.sebokwiki.org, p 688. Related Business Activities. (20 November 2023). *SEBoK*. Retrieved 14:55 UTC, December 23, 2023 from https://sebokwiki.org/w/index.php?title=Related_Business_Activities&oldid=70257
4. Wikipedia https://en.wikipedia.org/wiki/Enterprise_systems_engineering, Wikipedia:Text of the Creative Commons Attribution-ShareAlike 4.0 International License - Wikipedia
5. Cassone, Deandra T., Tillman, Frank A., *A Professional's Guide to Problem Solving with Decision Science: An Integrated Approach for Assessing Issues, Finding Solutions, and Reaching Corporate Objectives*, Second Edition Overland Park, KS, Pioneering Partnerships, LLC, January 2018
6. Tillman, Frank A., Cassone, Deandra T., *A Professional's Guide to Decision Science and Problem Solving: An Integrated Approach for Assessing Issues, Finding Solutions, and Reaching Corporate Objectives*, Upper Saddle River, NJ, Financial Times Press/Pearson Education, Web: December 1, 2012, Hardcover: February 2012

Team Effectiveness

Abstract

Engineers should expand beyond their technical skills to better understand administrative, management and team management functions in an organization. This includes understanding the difference between management and leadership, building high performance teams and a strong understanding of the human resource function. Areas such as these are often not innate knowledge areas for engineers and require the need for ongoing learning. This chapter addresses some high-level knowledge areas that should provide direction and important concepts to assist engineers in their career development efforts. Knowledge, skills and workstyles identified in the first chapter that are addressed in this chapter include the following:

Knowledge:

- Personnel and Human Resources
- Administrative

Expansion of core engineering characteristics at a broader organizational level:

- Judgment and Decision Making
- Coordination

Skills:

- Management of Personnel Resources
- Social Perceptiveness

D. Cassone, *Leadership Concepts for the Engineering Mindset*, Synthesis Lectures on Engineering, Science, and Technology, https://doi.org/10.1007/978-3-031-67798-4_7

Expanding Administrative, Management and Leadership Skills

To expand your leadership skills, you must understand how to work well with teams. Whether you are in a technical leadership position or managing the team, understanding how to interact with your team members will enable you to implement the design and development activities that you perform. You may not be the team manager or project manager, but if you don't interact well with your team members, it will be much harder to implement your ideas.

Many times, engineers learn management and leadership concepts the hard way. They would much rather be designing and developing solutions rather than managing people. This is understandable based on the engineer's mindset, but it is valuable for you to understand how you interact with your team and look for ways to improve those skills. This chapter is not intended to be exhaustive, but to point you to some concepts that will be helpful in your career. Keep reading and learning on this topic. It will only help you with your management and leadership development.

Management Versus Leadership

Management and leadership are similar but not the same. Managing is about being flexible and coping with complexity. Management addresses implementation, which is also important with leadership concepts, however, focuses more on the tactical execution of the vision of leadership. Management activities include:

- Organized and disciplined execution of tasks.
- Creates stability in the team by formulating plans, designing procedures, monitoring results against plans to detect problems and taking the necessary corrective actions to implement solutions.
- Finds a way to get things done and expedites activities.
- Makes tradeoffs among time, cost and schedule to implement solutions.
- Innovates and adapts to changing circumstances.
- Useful for well-defined projects.

Leadership builds on management skills to cope with project challenges and changes. Leadership sees the project and its requirements from an implementation perspective, but also approaches accomplishing tasks with a view of meeting the overall goals and expectations of the organization. Leadership activities include:

- Recognizes appropriate direction and vision.
- Articulates the direction and operation of the project.
- Aligns people with project direction.

- Motivates people to overcome obstacles.
- Addresses a higher uncertainty in projects in areas such as changes in budget and scope, technology issues, and the internal and external coordination of activity breakdowns.

A manager may have the opportunity to lead current and new projects. Understanding a broader view of your organization, cross-functional interactions, budgeting and other topics covered in this book expands your understanding so that you can integrate a more holistic view into the execution of your current tasks. This understanding of a more holistic view of your organization integrates more information into your engineering toolkit and expand your ability to solve problems from a bigger-picture perspective.

There is no substitute for leading by example. You must demonstrate your convictions by showing your team that you believe enough in them to do them yourself. Leadership is not based on professing an "edict on high" and having your "subjects" do what you say, not what you do. Personally demonstrating what you are asking your employees to do is one of the strongest tools that you have to motivate your team to do the same.

There is no substitute for leading by example. You must demonstrate your convictions by showing your team that you believe enough in them to do them yourself.

Leading by example is a highly visible, interactive management style and is essential to building cooperative relationships. You can utilize your own behavior as a powerful leadership tool and demonstrate how your team members should work. Below are some questions you may ask yourself to see how you are demonstrating your leadership to your team.

- **Priorities**—How do you spend your time to communicate the importance of an activity?
- **Urgency**—How do you approach deadlines, status reports, solution development and delivering results?
- **Problem solving**—Do you attack bad news, or do you take a proactive approach to developing solutions?
- **Cooperation**—How well do you cooperate with your team and how do you interact with leaders outside of your organization?
- **Performance**—What are your standards for delivering a quality project and how do you communicate those performance standards to your organization?
- **Ethics**—Are you ethical in your work or do you distort the truth? Your code of conduct can have serious consequences to your team and your organization.

Fundamentally, your team is watching you and your approach to leading the team. Leading by example is one of the most powerful tools to communicate expectations for your team.

Building Trust

Building trust with your team is essential. When issues arise, do you take responsibility as the manager or point fingers at your employees as the cause of the problem? When successes arise, do you take credit for the success or give that credit to your employees? Your employees are watching carefully how you handle these types of situations. You can be successful in the short term without trust, but a sense of mistrust damages your team's efficiency and performance in the long run. Ultimately, you in management are responsible for your team's performance, and managing with that in mind helps you to approach your team's behavior and performance.

Building trust with your team is essential. You can be successful in the short term without trust, but a sense of mistrust damages your team's efficiency and performance in the long run.

Distrust has inordinate amounts of time and energy spent discerning hidden agendas, true meanings of communications, hesitating to cooperate, and securing guarantees. Your team spends time deciphering meanings and actions instead of focusing on the work to be done. Trust maintains smooth and efficient interactions. With trust, others will take your intentions at face value instead of looking for hidden agendas. Trust helps also when there is organizational change and uncertainty.

Character focuses on personal motives. Deep rooted personal values are recognized by your team. Manipulation techniques and the "all about me" approach show employees that you will do what is best for you and may manipulate them to get what you want to succeed personally. To organically motivate your team, you will need to demonstrate positive character values and provide a sense of well-being and appreciation for the work your team does. Competence enables you to positively demonstrate character with your technical knowledge, listens and encourages and interacts positively with the organization at large.

High Performance Teams

A team is a group of people with complementary skills who work actively together to achieve a common purpose for which they hold themselves collectively responsible. Teams recommend specific solutions, set direction, develop strategies and persuade others. Teams design work units for project tasks with effective long-term relationships within membership, have a solid infrastructure of group processes and obtain external support to sustain energy, pace, and performance. Teams are a critical functional feature of accomplishing work for an organization.

What makes a team a high-performance team? Below are some attributes and characteristics of a high-performance team and ways that you can achieve this type of team [1–3].

- Common sense of purpose and willingness to work together to achieve project objectives.
- Identifies individual expertise, uses it appropriately and respects experts.
- Balances roles to facilitate task accomplishment and group cohesion.
- Exerts energy to solve problems rather than on interpersonal struggles.
- Encourages differences of opinion.
- Treats mistakes as opportunities for learning.
- Has high personal standards of performance.
- Encourages other team members to excel.
- Holds members identify with team and views the team as source of professional and personal growth.

Does your team exhibit these characteristics? Review the characteristics of a high-performance team listed above and assess how you may be able to better manage your team to exhibit the characteristics of a highly performing team.

A high preforming team can become one of your greatest assets. High performing teams become champions, create breakthrough products, exceed customer expectations, and get projects done ahead of schedule and under budget. They are bonded together by mutual interdependency and a common goal or vision. They trust each other and exhibit a high level of collaboration.

Human Resource Management

If you don't understand the importance of human resource management, you can end up in trouble with your team and the organization, overall. This is one area that you must know what is required from a legal and organizational perspective to manage and lead your team. Consulting your human resources department when you have questions about employees and employee issues can save you from making mistakes that could leave your organization vulnerable to issues and lawsuits.

One point to remember when dealing with human resources and your employees is that although human resources are there to help you and foster your career, they also function to protect your organization from legal issues. They will support you in your employee management functions, however, if you intentionally manage your employees unfairly, they will protect the organization from your actions. It is a bit of a harsh reality, but it is important to understand that the policies and procedures that they have in place are intended to not only protect the employee, but also to protect the organization.

Engineers can view themselves as engineers no matter what role they are performing in an organization. There can be a fundamental personal view that “I’m an engineer.” When you are in a management or leadership position, yes, you are still an engineer, but don’t forget that your job function is managerial and not engineering. Understanding human resource functions in a management position is required beyond your engineering degree. Unfortunately, some engineers learn this the hard way as did the author on one occasion. In dealing with an employee issue with human resources, the author made the comment, “but I’m an engineer.” The human resources team then said, “but your job is a manager.” Bingo! You only need to hear that one time to get the point, so pay attention to human resource requirements in your organization!

Typical fundamental responsibilities of human resource management are listed below [4].

- **Staffing**—This function includes human resource planning, screening, recruiting, and hiring.
- **Training and Development**—This function teaches organizational policies and procedures, employee skill training, employee professional development, and organizational change.
- **Motivation**—This function assesses the environment from the context of well-designed job functions. Additionally, it addresses performance appraisals, rewards and compensation and employee benefits.
- **Maintenance**—This function addresses the health and safety of the employee in the workplace. Additionally, this function keeps employees informed of current company activities and serves as an employee relations function.

Lack of knowledge of human resource management leads to a minefield of trouble. Below are some of the most common mistakes in human resource management. One small misstep by human resources or supervisors could cost your company thousands or even millions of dollars. Below lists some key mistakes in human resource management [5].

- **Hiring Process**—Avoid any discriminatory words in your position advertisements. Ensure that those interviewing the candidate do not ask about personal characteristics that could be viewed as discriminatory. Keep focused on the qualifications of the candidate being interviewed.
- **Wage and Compensation Issues**—Classify your employees correctly based on Fair Labor Standards as salaried or hourly employees. Apply overtime policies fairly and

correctly. Clearly specify work hours and location and any adaptations to their work schedules.

- **Training and Performance**—Ensure supervisors are adequately trained in how to conduct performance reviews and make wage determinations equitably and with appropriate rationale.
- **I-9 Form Mistakes**—Make sure all employees complete the appropriate I-9 forms and other necessary forms within the first three days of employment. Keep I-9 forms on file for at least three years after hiring and one year after termination.
- **Terminations**—Poorly handled terminations can result in lawsuits. Have another supervisor as a witness. Document the actions leading up to the termination and the termination meeting. Ensure that you are keeping an employee abreast of performance issues with regular feedback. Keep to the facts and avoid personalizing the termination. Don't discuss the reasons for termination with other employees.

Lack of knowledge of human resource management leads to a minefield of trouble.

Below are some tips for avoiding mistakes [6].

- **Identify Issues**—Identify the issues before they are brought to you by others.
- **Document, Document, Document**—Document extensively any disciplinary issues and performance discussions.
- **Establish Policies and Procedures**—Establish policies and procedures documented in an employee manual and ensure that your employees have been trained and acknowledge that they understand the policies. Provide supplemental training, as necessary.
- **Monitor Laws and Regulations**—Stay abreast of changing laws by monitoring updates to laws and regulations. Ensure that you are appropriately following the Fair Labor Standards Act in the classification of workers.
- **Hire and Fire Carefully**—Have a robust recruitment process, train your staff so that they can do their job well, and follow appropriate termination and pay processes.
- **Develop and Reward Employees**—Develop and reward employees by cross-training, providing development opportunities and appropriate rewards for good performance. Have formal performance reviews and periodic staff and one-on-one meetings with your employees.

Unfortunately, it may not be until you've had employee issues that you realize the importance of human resource management. Take time to keep yourself abreast of the needs of human resource management to avoid unnecessary employee issues.

Knowledge and Managerial Leadership

High Output Management, by Grove [7], and The Effective Executive, by Drucker [8], are valuable books for engineers and their career development. Both books recognize two different types of leaders in an organization, the managerial leader and the technical leader or knowledge/know-how leader. Not only do these books recognize the two different types of leaders, but they both provide tactical skills that can assist you in your career development.

From this author's perspective, the recognition of the two different types of leaders in an organization resonates heavily with engineers. Some engineers may not want to move into management positions but do want to be recognized for their technical contributions to an organization. Other engineers may want to move into engineering management positions and move beyond that to general management and executive leadership positions. Making a conscious decision with regards to your career is important.

Failing Forward Strategically

Many engineers are risk adverse, even if the risk may reap huge rewards. When presented with a risk with regards to a decision, design, etc., do so as a strategically calculated risk. Failures will happen, either personally or as a team, so develop an approach to assess and mitigate risks, learn the most that you possibly can from the failure and use it as a steppingstone to your advantage. These failures and how you and your team handle the failures can be a critical part of your career development. Failure one time may be the key to success in the next opportunity.

> **Failures will happen, either personally or as a team, so develop an approach to assess and mitigate risks, learn the most that you possibly can from the failure and use it as a steppingstone to your advantage.**

Leadership books such as one written by John Maxwell's "Fail Forward [9]," discuss the main reasons people fail, how to master fear and the positive benefits that can be derived from negative experiences. Failure is a circumstance, not a personal trait. It is a worthwhile book to read.

Below are some practical steps that you can take to enable you and your organization to fail forward [9].

- **Foster a Growth Mindset**—Position your organization from the perspective of growth. Learn from your team's positive and negative experiences and promote that mindset in your organization.
- **Create a Safe Environment For 'Failing Forward'**—Finger pointing, "throwing people under the bus" and other counteract positive learning from a failure experience.

Ensure that you are constructively addressing the components of the failure for positive use in the next experience.

- **Encourage Experimentation**—Some industry experts believe that for effective learning, leaders should plan to strategically produce experimental failures. Dissecting failures provides valuable information that can be leveraged for your next success.

Your behavior when your team fails is critical in demonstrating what risks your team can and cannot take. Avoiding risks may be "leaving money on the table" and avoid true progress. Be sure to understand your true perspective on risk and how you plan to address risks and failures with your organization. Strategically assessing your approach to risk management can be critical for you and your team's development. "The art of failing forward drives sustainability, and teaches collaboration, empowerment and resilience [9]."

Questions

Below are some questions to consider as part of this chapter.

- Can you point to individuals in your organization that are managers versus leaders? Why?
- Do you think that your team trusts you? Are there ways that you've lost their trust or ways that you can further build their trust?
- Does your team exhibit characteristics of a high-performance team?
- Are their ways that you can positively impact the performance of your team?
- Do you understand the role of human resources management? Have you taken courses offered by your company to improve your knowledge of needs for successful employee management?
- Are you in a "know-how" or management leadership position?
- How do you handle risk, failure, and success with your team?

References

1. Burkes, D. What Makes Some Teams High Performing?, hbr.org, August 30, 2023 What Makes Some Teams High Performing? (hbr.org)
2. Friedman, R. 5 Things High-Performing Teams Do Differently, hbr.org, October 21, 2021 5 Things High-Performing Teams Do Differently (hbr.org)
3. Keller, S., High-performing teams: A timeless leadership topic, McKinsey Quarterly, June 28, 2017, High-performing teams: A timeless leadership topic | McKinsey
4. Burokas, N., Functions of Human Resources Management, lumenlearning.com Functions of Human Resources Management | Human Resources Management (lumenlearning.com)

5. Business Management Daily, The 5 Most Dangerous Legal Mistakes HR Makes, http://www.businessmanagementdaily.com/37489/the-5-most-dangerous-legal-mistakes-hr-makes
6. Slater, T., 7 Costly HR Mistakes and How to Avoid Them, April 21, 2022, LinkedIn.com, 7 Costly HR Mistakes and How to Avoid Them (linkedin.com)
7. Grove, A., *High Output Management*, 2nd edition, 1995, Penguin Random House; New York, New York
8. Drucker, P., *The Effective Executive*, HarperCollins Publishers, New York, New York, January 6, 2006
9. Maxwell, J. "Fail Forward," HarperCollins Leadership; New York, New York, Reprint edition, April 1, 2007

Innovation—Engineering Entrepreneurship and Intrapreneurship

Abstract

Engineers are uniquely qualified to excel in technology development and innovation. Engineers may pursue commercializing their innovations in their own company as an entrepreneur or internal to an organization as an intrapreneur. Innovation leadership is a way for engineers to set themselves apart from others in the organization. Engineers should look for opportunities to innovate. Innovation can be useful to demonstrate leadership potential, both technically and managerially. Fundamental concepts and resources to support innovation are discussed in this chapter. Topics include the criticality of innovation, similarities and differences between entrepreneurs and intrapreneurs, business planning, pitch presentations and business financials. Additionally, a summary of a study used for innovation management by a large chemical company is discussed. Knowledge, skills and workstyles identified in the first chapter that are ad-dressed in this chapter include the following:

Knowledge:

- Administrative

Expansion of core engineering characteristics at a broader organizational level:

- Judgment and Decision Making
- Complex Problem Solving
- Critical Thinking

D. Cassone, *Leadership Concepts for the Engineering Mindset*, Synthesis Lectures on Engineering, Science, and Technology, https://doi.org/10.1007/978-3-031-67798-4_8

Skills:

- Management of Financial Resources
- Negotiation

Criticality of Innovation

Innovation in your work can set you apart from others in your organization. Innovation demonstrates that you not only understand the fundamentals of your discipline, but also demonstrates that you are able to apply your knowledge to improve products and operations. Innovation, in its simplest form is doing something better than was previously done. You most likely are presented frequently with a design, operation or implementation that can be done better. That's innovation. Use your critical thinking skills to assess the current state of operations and see if there is a better way to accomplish the same tasks.

Design, product and process improvements are a way to demonstrate your contribution as an engineer to your organization. A quantifiable measure of innovation is developing patentable designs and processes for your organization. Some organizations are focused on the development of patents from a strategic or protective perspective. A patent may seem to be a lofty goal but can encompass new and novel ways to do things and incremental improvements to products and processes. Check with your organization to see how they handle the patent process and investigate whether your ideas and improvements are patentable. Let the patent attorneys decide whether your ideas are patentable. You may be surprised what constitutes a patentable design or process. While working for a Fortune 100 company, the author had numerous patents awarded and was additionally given a company-wide innovation leadership award. Innovation can demonstrate your leadership and provide significant contributions to an organization.

Engineers are uniquely qualified to excel in technology and innovation. Your knowledge in technologies and abilities to problem solve bode well in the innovation space. Areas of innovation that you can contribute include the application of new technologies, use of new methods or tools, unique vendor engagement, application of best practices to your organization and many others. Stay abreast of new technologies and think about how new technologies can be applied to your workplace.

Engineers may see themselves from more of a domain application perspective, however, your problem solving, and critical thinking skills make you creative in your solution development. You can contribute creatively by demonstrating by applying fundamental engineering characteristics listed below.

- Producing ideas quickly.
- Generating original ideas.
- Perseverance in problem solving.
- Willingness to analyze and explore.
- Valuing intellectual contributions.

Characteristics of innovators may include the following.

- Strong technical background.
- Fluency in discussing and presenting new and novel ideas.
- Ability to analyze technical products.
- Belief that innovation is possible.
- Commitment to innovative concepts and products.
- Confidence in your innovation capabilities.

Engineering innovators may not realize the significance of their innovations. Innovation is important to your organization and can be very rewarding for you personally.

Engineering Entrepreneurship

Innovative engineers can jump from being an innovator within a company to becoming an engineering entrepreneur. Leadership concepts for engineers may transcend from a corporate leadership career path to that of a business owner and entrepreneur. This topic is covered in this book because it may be a natural leadership progression for some engineers. General familiarity with resources and concepts may provide a starting point to explore the entrepreneurship career option for engineers.

Numerous experts have written books, taught classes and provided short courses on engineering entrepreneurship. There are many great engineers to entrepreneur examples. If this is a topic of interest to you, begin by doing your homework and reading books to help you explore innovation and entrepreneurship. .A few fundamental books that may be helpful to you are included in the list below, however, this list is by no means all inclusive.

- The Lean Startup by Eric Ries
- The Innovator's Dilemma by Clayton Christensen
- Competing Against Luck by Clayton Christensen
- Business Model Generation by Osterwalder and Pigneur

This section of the book is not intended to teach you extensively about engineering entrepreneurship, however, to point out that it is a worthy concept to explore on your career path. Entrepreneurship in many ways represents a mindset. Can something be done better, faster, more customer centric and provide value to businesses and customers? Is it interesting and rewarding for you to explore and implement new concepts in your organization? Do you have ideas for a new product? Have you thought of a service model that will benefit your customers? Individuals with commitment and passion to answer those questions may accomplish this by founding a new business (entrepreneurship) or within an existing company (intrapreneurship). Engineering entrepreneurship and intrapreneurship is an engineering leadership concept.

Three crucial components of entrepreneurship are the opportunity, the customer need and timing as described below [1].

- **The Opportunity**—The biggest misconception about an idea for a new business is that it must be unique. Almost any idea you may have will also have occurred to others. What's important is developing the idea, implementing it, and building a successful business.
- **Customer Need**—You must be able to identify your prospective customers very specifically. There is no market unless customers have a real, proven need for the product. Customers must perceive that the new business will be giving them better value for their money than existing businesses do.
- **Timing**—Timing plays a crucial role in many opportunities. Windows of opportunity may only open once, or there may be steady demand for a product. If the window of opportunity is brief, it may be that the idea will quickly pass away too.

Regardless of how right the opportunity may seem to be, it will not make a successful business unless it is developed by a person with strong entrepreneurial and management skills. The idea itself is not enough. To build a business, you must have all the business components such as finance, marketing, distribution, etc. to bring your idea from conceptualization to actual customers. You must think of your entrepreneurial adventure from the big-picture perspective.

Great resources exist for engineering entrepreneurs whether it be books, classes, mentors, and other organizations to assist entrepreneurs with their business ideas. It is beneficial for an engineer to have an entrepreneurial mindset that can lead to very rewarding career opportunities within and outside of your current organization.

It is beneficial for an engineer to have an entrepreneurial mindset that can lead to very rewarding career opportunities within and outside of your current organization.

The author teaches a engineering entrepreneurship class in a graduate engineering management program. Engineers from all engineering disciplines enroll in the class. The breadth of ideas that are explored by engineers for developing a business is interesting. Some students have developed business plans for ideas such as a RV-based AirBnb in the Flint Hills, manufactured gaming components, pedal brew pubs, a cider production business, a drone photography business, consulting businesses and more. It's quite interesting to see the breadth of ideas that these engineers are looking to pursue. Most of these engineers, also, are evolving into leaders in their own organizations.

Engineering Entrepreneurship Versus Intrapreneurship

Engineers are positioned to innovate. How best can you as an engineer utilize those skills within or outside of your organization? Two ways that you can pursue the path of innovation is to innovate outside of your current organization (entrepreneurship) or innovate within your organization (intrapreneurship). Both avenues allow you to design, develop and launch new or improved products and services. Engineering Intrapreneurship may provide you with a testing ground for ultimate engineering entrepreneurship.

There are similarities and differences between an intrapreneur and an entrepreneur. You should understand the operational differences between the two types of innovators so that you can set your expectations appropriately. Entrepreneurs notice opportunities and take the initiative to mobilize resources to make new goods and services. Intrapreneurs also notice opportunities and take initiative to mobilize resources, however they work in large companies and contribute to company innovation. Intrapreneurs often become entrepreneurs. Learning organizations encourage intrapreneurship. Intrapreneurs take ownership of a product from concept to market.

Leaders try to foster entrepreneurship inside the organization, but inadequate attention is given to the differences. Entrepreneurs willingly sign up for a wide spectrum of possible outcomes, ranging from big wins to massive losses. Intrapreneurs have made the conscious decision to receive a paycheck from an organization and don't expect huge financial gains. Entrepreneurs have the luxury of building a company's culture from scratch. However, intrapreneurs often fight against deeply held and widely shared beliefs. Intrapreneurs are not as autonomous as entrepreneurs, although they are typically made to believe they are autonomous which can lead to disillusionment. Entrepreneurs directly hire marketing, advertising, research and development, design, engineering, and sales talent. Intrapreneurs must negotiate for these and other scarce resources, subject to availability and competing projects. Most people believe intrapreneurs are simply internal entrepreneurs who can be expected to act in the same ways as entrepreneurs, however, in practice, this is not the case [2].

Leaders try to foster entrepreneurship inside the organization, but inadequate attention is given to the differences.

The benefits of acting as an intrapreneur within your organization stem from receiving a steady paycheck. Entrepreneurs are not guaranteed consistent compensation, however, can have significant reward. One key benefit of acting as an intrapreneur is to test your ideas in a safer concept and critically think about what would be needed for you to become an entrepreneur. It can be a great learning environment and provide you with additional opportunities as an innovator in your organization or provide you with additional knowledge to eventually go out on your own as an entrepreneur.

Planning and Pitching Your Ideas

Whether you are functioning as an entrepreneur or an intrapreneur, you must plan. From the perspective of an entrepreneur, you must develop a business plan to document your idea and test the financial success of your endeavor. From an intrapreneur's perspective, you will need to develop a business case as per organizational formats to receive internal funding for your effort. Within an organization, this may be formal or a more informal process, but having the business case provides you with quantifiable planning facts that you can use to communicate. A typical business plan incorporates key elements such as an executive summary, industry, customer and competitor analysis, company and product description, marketing plan, development plan, team, crucial risks, offering and a financial plan. An internal business case many not required the same level of detail as an entrepreneur's business plan.

Developing a business plan or business case can be a significant effort. This exercise requires you to think through each key element of your business and describe how you intend to accomplish your plan. From an entrepreneur's perspective, this document can be used to obtain funding and communicate your idea (include confidentiality statements on the plan if you do share it). It also forces you to think about areas that you may not have explicitly thought about with regards to the industry analysis, marketing plan, operations plan and critical risks [3]. As an intrapreneur you may not be required to have the same level of rigor, but it is still good to consider the key elements of how your innovation integrates within your organization.

You also need to be able to communicate your plan clearly and succinctly in a short period of time. Think "Shark Tank." You may not have much more time than a few minutes to interest potential investors or stakeholders in your idea, internally and externally. Spending time developing a summary of your plan that you can communicate quickly is very important. Many times, this is called a Pitch Presentation.

Key elements of a Pitch Presentation from Guy Kawasaki's *The Art of the Start* summarize the key information that should be included in the initial pitch deck [4]. Key slides that he recommends for pitch presentations include the title slide with contact information, problem you are alleviating, solution to the problem, the business model and you intend to make money, the underlying magic of the technology and enhancements, market analysis, competition, management team and financial projections. Spend a significant amount of time developing this presentation. Be clear and succinct in your slides and presentation. This is an important mechanism to communicate your idea so you want to demonstrate competence, the ability to succeed, and will prudently use funding you may get from the investors. You are selling yourself as well as your business idea.

Your financial plan is critical, whether you are an entrepreneur or functioning as an intrapreneur within your organization. You must be able to support your business idea financially. Below are some key drivers of your financial plan.

- **Revenue**—How do you plan on generating revenue for your business? Do you have a single or multiple income stream? What is the turn-around time from revenue generation to payment?
- **Gross Profit Margin**—Will you offer discounts on your product for bulk sales? How expensive are your employees? Is there material waste in your production? What are your fixed and variable operating costs?
- **Large Operating Expenses**—Do you need large facility or equipment cost to support your product? What are advertising costs? Will you incur large research and development costs? Are there licenses you must obtain?
- **Customer Acquisition Costs**—How much will it cost you in marketing to gain a new customer? What are your projected marketing and sales expenses?
- **Cash Flow**—Positive cash flow is critical to the sustainability of a company. What are your working capital assumptions? Do you have grants to support your start-up? What are your capital expenditures and depreciation, tax regulations and rates, etc.?

Your financial plan is critical, whether you are an entrepreneur or functioning as an intrapreneur within your organization. You must be able to support your business idea financially.

You should develop financial statements such as current or forecast income statements, balance sheets, statement of cash flows and retained earnings on your key drivers over a time horizon of three to five years. You want to demonstrate that you have the right team in place to succeed and have thoroughly projected the financials for your organization. These financial statements along with the development of your business plan benefit you as well as investors to decide whether this is a viable endeavor.

An internal corporate business case used by an intrapreneur most likely will not require the rigor that is required for an entrepreneur to develop a business plan. You will, however, want to provide a thorough analysis of the financials over time and incorporate the time value of money in the financial computations. Understanding and communicating the impacts on customers, competition, market, operations, risks and the overall financial plan will be important to obtaining the internal resources necessary to support your product or improvement.

Innovation Management

Most organizations are committed to innovation, they launch products to sustain their product base and gain market share. Innovation and new product development keeps an organization current and alive. How does your organization assess, make interim decisions to proceed and fund a portfolio of new product development? Organizations use a spectrum of techniques such as committee decision making and structured decision processes to manage their portfolio of products from conceptualization through product launch. The following study demonstrates one approach to structured decision making for new product development processes. This study and decision support system was developed for new product development in the early 1990s for a Fortune 100 chemical company [5].

Study: Strategic Planning Model for New Product Development

This study describes a strategic planning model developed for selecting new products for the future of the company. This application provides a good example of an automated decision model for new product development and corporate planning. The application was developed to systematically manage and evaluate business and/or new product opportunities from conception to commercialization [6].

In the development of the model, decision makers from various levels of the organization established the goals and objectives of the organization. These goals and objectives were then translated into decision criteria and metrics that could be used to measure the success of each product development effort in meeting the organizational goals.

The decision tool was created to evaluate development programs through five stages of implementation. Several criteria were scored for each program. The criteria were then weighted and used to rank all the development programs. The criteria included the following:

- Market—type of market and potential sales volume
- Finance—net present value, cash flow streams, life cycle, and payout
- Resources—what physical resources are required, for example, building, equipment, and staffing
- Probability of success—1–100% estimated probability of meeting all expectations
- Strategic alignment—Does this prospective opportunity complement current business activities and are there any barriers (patents), protection, or technology that the organization possesses to ensure success?

Multiple Criteria Ranking Methodology

The ranking used a prioritization methodology that incorporated both subjective and quantitative inputs. This methodology, TOPSIS (Technique for Order Preference by Similarity to Ideal Solution) [6], was based on the relationships between the criteria and their values and weights. The ranking model was a hierarchical design: most of the top-level criteria have sub-criteria. For example, the financial criteria are the Year of the First Revenue Flow, Number of Years to Last Revenue Flow, and Net Present Value.

Sensitivity Analysis

The decision support system calculated the sensitivity of each criterion input. The system calculated the amount each criterion must change to move that specific program up one rank or down one rank on the list. Some criteria were more sensitive than others and had a greater impact if changed.

A graphing section was included that graphed the Net Present Value against the Probability of Commercial Success, which showed the most profitable combination of programs for the business.

Stage-Gate Process

This application focused on the use of a stage-gate process. The goal was to find the most feasible and profitable investment alternatives early in the development process and to eliminate investments at their earliest stages that appeared to be less desirable to the company's overall financial and strategic objectives. The systematic evaluation process using the multiple criteria methods coupled with the stage-gate evaluations provided a basis for the holistic alignment with the company's overall objectives.

The Decision Support System

The system included options to view the program inputs and the ranked program list, print standard reports, and save or retrieve the data for transfer between systems. The system was built as an evaluation tool at each stage in the five stage-gate process from program conception to commercialization.

At each gate in the process, the programs were evaluated using different criteria and different weights on the criteria. The gate authorization of each program was set as the program progressed through the development stages. The system provided for the following:

- Tracked developmental progress of current business and/or new product opportunities.
- Eliminated unprofitable business and/or new products at the earliest point in the developmental phases.
- Generated a prioritized business and/or new product opportunity listing.
- Utilized discipline and consistency in evaluating each business and/or new product opportunity.

- Calculated the sensitivity of each criteria value to determine the impact of a change in the data.
- A decision tool, which used the stage-gate process to evaluate the business or new product at each developmental step, standardizes the evaluation process of the new products.

When investing millions of dollars in innovation and new product development, it is important to have a structured way to evaluate your opportunities and assess their contributions to organizational goals. The structured process itself will assist you in critically assessing your investments as they contribute to organizational objectives and enable effective resource utilization.

Does your organization have a structured process to manage innovation and new product development? Understanding the organization's goals and metrics along with the financial and implementation processes is critical to the effective use of resources in launching innovation. Structured decision making and innovation leadership is required to effectively use an organization's limited resources.

When investing millions of dollars in new product development, it is important to have a structured way to evaluate your opportunities and assess their contributions to organizational goals.

Questions

Below are some questions to consider as part of this chapter.

- Do you see entrepreneurship or intrapreneurship in your career path?
- Have you read books, taken classes, or talked to individuals about your ideas?
- Have you begun the planning process to be an entrepreneur?
- Do you have the opportunity to function in a position as an intrapreneur within your organization?
- Are you managing innovation in your organization? Do you have a structured process to do so?

References

1. Bygraves, W. D, Zacharakis, A., "*The Portable MBA in Entrepreneurship*" Fourth Edition, Wiley and Sons, November 16, 2009, Hoboken, New Jersey
2. Myler, L., Intrapreneurs Are Just Like Entrepreneurs...NOT!, Forbes, Jan 15, 2014, http://www.forbes.com/sites/larrymyler/2014/01/15/intrapreneurs-are-just-like-entrepreneurs-not/#31b39f1e165c
3. Lawrence, S., Moyes, F., Business Plan Preparation - Tools for writing business plans, Deming Center for Entrepreneurship, Graduate School of Business, University of Colorado, Boulder, Colorado 80309-0419 Business Plan Templates (colorado.edu) Elements of a Successful Business Plan (colorado.edu)
4. *Kawasaki, G., The Art of the Start: The Time-Tested, Battle-Hardened Guide for Anyone Starting Anything, 2004, Toronto: Penguin Canada.* (https://www.marsdd.com/mars-library/how-to-create-a-pitch-deck-for-investors/)
5. Section Reprinted, Tillman, Frank A., Cassone, Deandra T., Strategic Planning and New Product Development, Financial Times Press/Pearson Education, Upper Saddle River, NJ, Web. May 31, 2012
6. Hwang, C.L.; Yoon, K. (1981). *Multiple Attribute Decision Making: Methods and Applications.* New York: Springer

The Rest Is Up to You

Abstract

Engineering leadership characteristics may present themselves to engineers early in life. This chapter discusses a write-up of an engineering leader written at the time he was thirteen years old. The write-up demonstrates the engineering mindset at an early age and this chapter discusses how those engineering mindset characteristics translate into engineering leadership positions later in his career. In many ways, it summarizes a key point in this book; Engineers have innate knowledge and skills that translate from early career engineering qualities to corporate leadership and business ownership. You, too, should look for those same examples and characteristics in your own career development and evolve that assessment to your current career. Knowledge, skills and workstyles identified in the first chapter that are ad-dressed in this chapter include the following:

Knowledge:

- Personnel and Human Resources
- Administrative
- Economics and Accounting
- Sales and Marketing

Expansion of core engineering characteristics at a broader organizational level:

- Judgment and Decision Making
- Complex Problem Solving
- Critical Thinking
- Coordination
- System Evaluation

D. Cassone, *Leadership Concepts for the Engineering Mindset*, Synthesis Lectures on Engineering, Science, and Technology, https://doi.org/10.1007/978-3-031-67798-4_9

Skills:

- Management of Personnel Resources
- Social Perceptiveness
- Management of Financial Resources
- Negotiation

The Engineering Mindset

Studying and observing what makes engineers successful has always been of interest. Engineers' logical, critical thinking and problem-solving approaches can lead to great successes in leadership and innovation for an engineer. As an engineer progresses in their career, they need to expand those key skills to broader organizational functions to be successful. Without an expansion of the breadth of knowledge of an engineer to a holistic organizational view, an engineer may not function successfully in managerial and leadership positions. Fundamentally, what do those engineers that end up in corporate leadership positions do differently than those that may aspire to those positions but do not reach them?

There are numerous examples of engineers that have successfully climbed the organizational leadership ladder. These individuals may end up as senior executives in corporations, university presidents and senior leadership or technical experts that shape industries and research. There is one engineer with great personal familiarity that helps trace the leverage of the engineering mindset in organizational and personal success and is discussed in this chapter.

Frank A. Tillman, Ph.D., PE was an engineering leader born with an engineering mindset. Frank came from a family of seven children whose father died when he was nine years old which necessitated him beginning to work at time. Both academically and athletically talented, he first attended college on a basketball scholarship, but decided to focus his efforts academically. He married in his early twenties and worked full time to put himself through college and graduate school and support his family. After a four-year job as an operations research analyst at a major oil company, he went on to pursue his Ph.D. full time. After graduating, he was in his first faculty position for a year and a half and then was named an engineering department head at a state land grant university at twenty-nine. Frank welcomed the leadership role and department head position and developed a robust engineering program in his twenty-one-year full-time tenure at the university. Frank then had the opportunity to launch and run two consulting companies that provided consulting services to government, military, and commercial clients. Additionally, he was a real estate developer building numerous housing developments and commercial real estate projects. He supported his community by participating in advisory councils, was on the local school board for eight years and coached numerous basketball, softball, and baseball teams. He mentored many successful executives and was emphatically committed to encouraging students to pursue engineering degrees. Frank was also extremely committed to his family and spent a significant amount of time with them. Frank Tillman was an engineering leader on steroids!

The below text was written by Frank at the time that he was about thirteen years old. His engineering mindset was evident at that early age. The write-up is about "Training for a Baseball Game." The concepts that are demonstrated in the write-up relate back to many of the topics that are addressed in this book. It is interesting to know of the success of Frank and to see how he analyzed and solved problems at an early age. Many of you have a similar mindset, and, as you can see, it can be used successfully in many different areas. As an engineer, you most likely have experiences and examples that have led you to becoming an engineer [1].

Training for a Baseball Game

To train for a baseball game or another even [another game] for that matter, you have to learn the rules. This little detail has caused so many disappointments that it would be worth your while to learn the rules. Study the latest rule book for any recent changes in the rules and only after you have caught up on the latest, should you start to train.

Obtain a good coach who can spot defects and effects quickly and can give the best remedies for them.

After getting a coach start looking for some place to play. A good field is very important in having a good ball game. Have the diamond fixed up so there are no bumps as bumps that would obstruct the path of the ball. If there are small rocks and chips and so forth in the field have it raked. Then after that drag the field so that it will leave a fertile top instead of a hard crust to play on. The dragging should be done in the field only. After the field and coach have been taken care of, equipment is the next consideration.

The catcher has to have a mask and shirt protector as well as shin guards. Besides the catcher no other player needs any special equipment except the usual stuff of course and that is spikes, uniform and mitt.

With all the preparation taken care of we are now ready to start the real training. Then first of all the whole team is sent out on the playing field for the purpose of seeing what each individual player was capable of doing. And from this check improvements are made. Before practice begins the coach gets the players out on the field for some muscle loosening exercises to loosen up all the tight muscles that might hinder the performance of a player making a difficult play.

The hitting is one of the most important factors of a good baseball team. As each player takes his turn at batting practice the coach watches his every movement to find some flaw or imperfection which lessens the power behind the ball. If he spots one he will have the batter correct it.

When batting practice is over the coach then has the fellows practice base running. Right after base running sliding is next, but before we go into that the coach gives pointers to the runners on how to get off with a fast start and how to avoid taking any unnecessary steps while going around the bases. After this he explains the proper method of sliding.

Sliding is a very essential part in baseball stealing because when you slide you are close to the ground and the baseman has to bend down to tag you thus giving you more time to be safe.

After learning to hit, run and steal we now go to the field. Fielding the ball so as to get the runner takes great skill and much practice. This is the part of practice the coach drills you most on. He sees you catch a few and then points out various mistakes that you should check yourself on. Then he gives you a few good pointers and sends you back out into the field.

The rest is up to you. End.

Skills of an Engineer and a CEO

Below are the top skills of an Engineer as discussed in the first chapter. The write-up in the previous section shows early evidence of Frank with engineering leadership skills.

Engineering Discipline Skills

- **Complex Problem Solving**—Assessing the overall problem. Breaking down the problem into specific areas of focus.
- **Critical Thinking**—Understanding rules as the framework for playing baseball. Addressing each area and discussing the requirements and needs for each. Understanding the need for a coach to point out flaws.
- **Mathematics**—Identifying the diamond for the field. Discussing minimizing the effort when running. Using sliding to your advantage.
- **Reading Comprehension**—Reading the rules.
- **Active Learning**—Being sure you are up to date on the new rules. Learning from the coach. Understanding the need for practice.
- **Active Listening**—Listening to the coach so that you can improve.
- **Operations Analysis**—Assessing the big picture of a baseball game and its components.
- **Science**—Warming up muscles for optimal performance.
- **Speaking**—Communicating your thoughts, verbally and on paper.
- **Systems Analysis**—Breaking down baseball into its components to optimize overall performance.
- **Judgment and Decision Making**—Knowing that you need to learn all the components of baseball to be successful. Practice in each area and focus on fielding.
- **Monitoring**—Adjusting when flaws are identified.

- **Systems Evaluation**—Assessing baseball from an overall perspective and breaking it down into parts for specific areas of mastery.
- **Coordination**—Bringing all the components of baseball together to be successful.
- **Time Management**—Focusing on specific skills that will bring you better results.
- **Writing**—The ability to write well for a thirteen-year-old.

Below are the top ten skills of a CEO that are discussed in the first chapter. You can see how Frank evolved his early skills to successfully run an engineering department, two consulting companies, develop and manage real estate, support the community and support his family.

Chief Executives

- **Judgment and Decision Making**—Able to manage different companies in different operating domains.
- **Complex Problem Solving**—Solve complex business management decisions across multiple domains and solve complex problems for clients.
- **Critical Thinking**—Able to assess each company and activity and successfully manage each.
- **Coordination**—Coordinate activities within a company and between various companies.
- **Management of Financial Resources**—Manage the financials and resources for consulting companies and real estate businesses.
- **Management of Personnel Resources**—Manage staff, consultants, technical employees, and relationships in the academic, business and community domains.
- **Speaking**—Exceptional at delivering presentations and assessing problems "on his feet."
- **Systems Evaluation**—Evaluate corporate, military and government systems and develop solutions for clients. Assess the operating systems of his companies.
- **Negotiation**—Negotiate with clients to generate contracts and problem solutions.
- **Reading Comprehension**—Avid reader with interests in engineering, management, finance, accounting, history, and politics.

Frank did many things to develop his skills as an engineer and move to higher levels of responsibility. In high school, along with working to support his single mother and family, he played sports and read every book in the high school library. In college, he took up to twenty-two credit hours a semester at a time that you did not need to pay for courses above 15 hours, so he could learn this additional information "for free." While working for Standard Oil of Ohio, he pursued his Ph.D. and finished that degree at the University of Iowa with a prestigious Ford Foundation scholarship. As a professor, he took on the responsibility of being a department head at an early age. At that time, he also started his first housing development, learning as he went to understand the requirements for housing infrastructure, marketing and sales. The list goes on, however, fundamentally, he continually learned and optimistically took on additional responsibilities in his career.

Frank was also very committed to student education and mentoring. He mentored many successful executives and took his responsibilities for educating engineering students very seriously. His time on the local school board was important to him in contributing to primary and secondary school education. Anyone that knew him knew his commitment to education and his belief that engineers had the characteristics and capabilities to be leaders.

Your Engineering Mindset

Perform a similar exercise yourself. Can you identify your early skills, current knowledge and skills as an engineer and then those that you see yourself performing at higher levels of responsibility in an organization? Look at what you think that you might need to achieve your career goals. Do you need to take additional micro credential or university courses? Do you need additional experience in your organization, or can you take that on in a community organization to gain leadership skills? Do you read books that will assist you in your development? Are there mentors that you can identify and talk with to determine how they have achieved the position that they are currently holding? Take time to evaluate your career and what knowledge and experience you need to move forward with your career development.

As an engineer, you can evolve your career to those positions best suited for you. Not everyone wants to be an executive or a CEO, and it is not necessary to be in those positions to have a successful career. Be happy and do the things that interest you most. The suggestions in this book are provided for you to look at your current career, assess skills that you may lack, develop new skills and make decisions on your career path. No one can make those choices for you, but the suggestion is to actively assess your choices and pursue them in a way that is best for you. Many, many resources exist and pursue those areas of interest to find your best career fit. As Frank said many times throughout his life, "The rest is up to you."

Questions

Below are some questions to consider as part of this chapter.

- When did you decide to become an engineer?
- Do you have examples of your thinking process at an early age?
- Can you see how your skills as an engineer can translate to your selected career path?

- Can you see how you can translate your engineering approach to other positions in an organization?
- Have you identified skills and knowledge you may need to support your career path decisions and how you can gain those skills and knowledge?

Reference

1. Tillman, Frank A., *Training for a Baseball Game*, written in approximately 1950

Appendix

Detailed O*NET Analysis

This appendix provides supporting details with regards to the discussion of Knowledge, Skills and Workstyles described in the O*NET program discussed in this book. Additional details are available on the O*NET website along with information with regards to other O*NET content model and other engineering disciplines. There is some insightful information in the O*NET system and would be useful for further study.

Table A.1 shows the five engineering disciplines reviewed in this assessment. Included in the chart are job descriptions and a sample of the reported job titles for each discipline. This information is provided for context and reference as there is a closer look at the skills and knowledge that are required for each position.

Table A.2 shows for management and leadership positions selected for the analysis. This table provides context for the job descriptions and sample titles that these positions entail.

Knowledge

As described by O*NET, Knowledge is an "Organized sets of principles and facts applying in general domains" [1]. These domains represent key information that an individual in the position should have. Table A.3 shows knowledge areas that the five selected engineering disciplines should possess.

From a management perspective, below are the knowledge areas that an engineering manager, general manager and executive should possess as shown in Table A.4.

The following list shows each of these knowledge areas. See Appendix "Definition of O*NET Knowledge Areas, Skill Areas and Work Styles" for a full definition of the knowledge areas.

D. Cassone, *Leadership Concepts for the Engineering Mindset*, Synthesis Lectures on Engineering, Science, and Technology, https://doi.org/10.1007/978-3-031-67798-4

Table A.1 Five engineering disciplines [1]

Disciplines	Job description	Sample of reported job titles
Mechanical engineers	Perform engineering duties in planning and designing tools, engines, machines, and other mechanically functioning equipment. Oversee installation, operation, maintenance, and repair of equipment such as centralized heat, gas, water, and steam systems	Application engineer, design engineer, design maintenance engineer, equipment engineer, mechanical design engineer, mechanical engineer, process engineer, product engineer, project engineer, test engineer
Electrical engineers	Research, design, develop, test, or supervise the manufacturing and installation of electrical equipment, components, or systems for commercial, industrial, military, or scientific use	Circuits engineer, design engineer, electrical controls engineer, electrical design engineer, electrical engineer, electrical project engineer, engineer, instrumentation and electrical reliability engineer (I&E reliability engineer), project engineer, test engineer
Chemical engineers	Design chemical plant equipment and devise processes for manufacturing chemicals and products, such as gasoline, synthetic rubber, plastics, detergents, cement, paper, and pulp, by applying principles and technology of chemistry, physics, and engineering	Chemical engineer, development engineer, engineer, engineering scientist, process control engineer, process engineer, project engineer, refinery process engineer, research chemical engineer, scientist
Civil engineers	Perform engineering duties in planning, designing, and overseeing construction and maintenance of building structures and facilities, such as roads, railroads, airports, bridges, harbors, channels, dams, irrigation projects, pipelines, power plants, and water and sewage systems	City engineer, civil engineer, county engineer, design engineer, engineer, geotechnical engineer, licensed engineer, project engineer, railroad design consultant, structural engineer
Industrial engineers	Design, develop, test, and evaluate integrated systems for managing industrial production processes, including human work factors, quality control, inventory control, logistics and material flow, cost analysis, and production coordination	Continuous improvement engineer, engineer, facilities engineer, industrial engineer, operations engineer, plant engineer, process engineer, project engineer, quality engineer, research and development engineer (R and D engineer)

Table A.2 Management and leadership positions [1]

Disciplines	Job description	Sample of reported job titles
Architectural and engineering managers	Plan, direct, or coordinate activities in such fields as architecture and engineering or research and development in these fields	Civil engineering manager, electrical engineering manager, engineering director, engineering group manager, engineering program manager, mechanical engineering manager, process engineering manager, project engineering manager, project manager
General and operations managers	Plan, direct, or coordinate the operations of public or private sector organizations, overseeing multiple departments or locations. Duties and responsibilities include formulating policies, managing daily operations, and planning the use of materials and human resources, but are too diverse and general in nature to be classified in any one functional area of management or administration, such as personnel, purchasing, or administrative services. Usually manage through subordinate supervisors. Excludes first-line supervisors	Business manager, center manager, department manager, district manager, general manager (GM), operations director, operations general manager (operations GM), operations manager, plant superintendent, store manager
Chief executives	Determine and formulate policies and provide overall direction of companies or private and public sector organizations within guidelines set up by a board of directors or similar governing body. Plan, direct, or coordinate operational activities at the highest level of management with the help of subordinate executives and staff managers	CEO (chief executive officer), chief diversity officer (CDO), chief financial officer (CFO), chief information officer (CIO), chief operating officer (COO), chief technical officer (CTO), executive director, executive vice president (EVP), operations vice president (operations VP), president

Table A.3 Knowledge areas of engineering disciplines [1]

Mechanical engineers	Electrical engineers	Chemical engineers	Civil engineers	Industrial engineers
Engineering and technology	Engineering and technology	Engineering and technology	Design	Engineering and technology
Mathematics	Computers and electronics	Chemistry	Engineering and technology	Production and processing
Design	Design	Mathematics	Building and construction	Mechanical
Mechanical	English language	Physics	Mathematics	Design
Physics	Mathematics	Design	English language	English language
English language	Physics	Production and processing	Physics	Mathematics
Computers and electronics	Production and processing	English language	Administration and management	Computers and electronics
Production and processing	Mechanical	Computers and electronics	Customer and personal service	Administration and management
Customer and personal service	Public safety and security	Public safety and security	Law and government	Customer and personal service
Chemistry	Customer and personal service	Mechanical	Public safety and security	Education and training

Table A.4 Knowledge areas of management and leadership positions [1]

Architectural and engineering managers	General and operations manager	Chief executives
Design	Administration and management	Administration and management
Engineering and technology	Customer and personal service	Personnel and human resources
Mathematics	English language	Customer and personal service
Administration and management	Production and processing	English language
English language	Mathematics	Economics and accounting
Customer and personal service	Personnel and human resources	Public safety and security
Mechanical	Administrative	Computers and electronics
Computers and electronics	Economics and accounting	Sales and marketing
Physics	Sales and marketing	Mathematics
Production and processing	NA	Education and training

- Administration and Management
- Administrative
- Building and Construction
- Chemistry
- Computers and Electronics
- Customer and Personal Service
- Design
- Economics and Accounting
- Education and Training
- Engineering and Technology
- English Language
- Law and Government
- Mathematics
- Mechanical
- Personnel and Human Resources
- Physics
- Production and Processing
- Public Safety and Security
- Sales and Marketing

Figure A.1 shows a comparison of the base engineering disciplines, how often knowledge areas were identified across the engineering disciplines and knowledge areas required by management and executive positions. This figure is similar to Fig. 5 in the text, but also shows the count that the various knowledge areas were listed for the five engineering disciplines analyzed. Areas highlighted in orange represent engineering knowledge areas. Areas highlighted in green represent management skills that aren't typically possessed by engineers from a functional engineering perspective. This provides some quantitative data to show engineers areas that they may need additional knowledge, skills and training.

Skills

As described by O*NET, "Experience lays the foundation for establishing procedures to work with given knowledge. These procedures are more commonly known as skills" [1]. Skills are used in day-to-day work, and it is important to understand what skills engineers traditionally may possess and those skills that they may need to attain in their career development. Table A.5 shows skill areas that the five selected engineering disciplines should possess.

From a management perspective, Table A.6 shows the skill areas that an engineering manager, general manager and executive should possess.

The following list shows each of these skill areas. See Appendix "Definition of O*NET Knowledge Areas, Skill Areas" for a full definition of the skill areas.

Knowledge Area Across All Engineering Disciplines	Number of Engineering Disciplines
Design	5
Engineering and Technology	5
English Language	5
Computers and Electronics	4
Customer and Personal Service	4
Mathematics	4
Mechanical	4
Physics	4
Production and Processing	4
Public Safety and Security	3
Administration and Management	2
Chemistry	2
Building and Construction	1
Education and Training	1
Law and Government	1

Architectural and Engineering Managers
Design
Engineering and Technology
Mathematics
Administration and Management
English Language
Customer and Personal Service
Mechanical
Computers and Electronics
Physics
Production and Processing

General and Operations Managers	Chief Executives
Administration and Management	Administration and Management
Customer and Personal Service	Personnel and Human Resources
English Language	Customer and Personal Service
Production and Processing	English Language
Mathematics	Economics and Accounting
Personnel and Human Resources	Public Safety and Security
Administrative	Computers and Electronics
Economics and Accounting	Sales and Marketing
Sales and Marketing	Mathematics
Not Listed on O*NET	Education and Training

Legend
Engineers
General Management

Fig. A.1 Comparison of knowledge areas between disciplines including counts [1]

- Active Learning
- Active Listening
- Complex Problem Solving
- Coordination
- Critical Thinking
- Judgment and Decision Making
- Management of Financial Resources
- Management of Personnel Resources
- Mathematics
- Monitoring
- Negotiation
- Operations Analysis
- Reading Comprehension
- Science

Table A.5 Skill areas of engineering disciplines [1]

Mechanical engineers	Electrical engineers	Chemical engineers	Civil engineers	Industrial engineers
Active listening	Writing	Science	Active listening	Active listening
Critical thinking	Critical thinking	Complex problem solving	Complex problem solving	Critical thinking
Mathematics	Reading comprehension	Critical thinking	Critical thinking	Reading comprehension
Reading comprehension	Active learning	Judgment and decision making	Mathematics	Complex problem solving
Science	Active listening	Mathematics	Reading comprehension	Speaking
Complex problem solving	Complex problem solving	Reading comprehension	Speaking	Writing
Judgment and decision making	Speaking	Systems analysis	Science	Monitoring
Operations analysis	Monitoring	Active learning	Systems analysis	Active learning
Active learning	Mathematics	Systems evaluation	Time management	Mathematics
Systems evaluation	Coordination	Operations analysis	Operations analysis	Systems analysis

Table A.6 Skill areas of management and leadership positions [1]

Architectural and engineering managers	General and operations managers	Chief executives
Reading comprehension	Active listening	Judgment and decision making
Active listening	Monitoring	Complex problem solving
Complex problem solving	Reading comprehension	Critical thinking
Speaking	Speaking	Coordination
Writing	Coordination	Management of financial resources
Active learning	Critical thinking	Management of personnel resources
Critical thinking	Management of personnel resources	Speaking
Judgment and decision making	Social perceptiveness	Systems evaluation
Mathematics	Active learning	Negotiation
Time management	Complex problem solving	Reading comprehension

Engineering Discipline Skills	Number of Engineering Disciplines
Complex Problem Solving	5
Critical Thinking	5
Mathematics	5
Reading Comprehension	5
Active Learning	4
Active Listening	4
Operations Analysis	3
Science	3
Speaking	3
Systems Analysis	3
Judgment and Decision Making	2
Monitoring	2
Systems Evaluation	2
Coordination	1
Time Management	1
Writing	1

Architectural and Engineering Managers
Reading Comprehension
Active Listening
Complex Problem Solving
Speaking
Writing
Active Learning
Critical Thinking
Judgment and Decision Making
Mathematics
Time Management

General and Operations Managers	Chief Executives
Active Listening	Judgment and Decision Making
Monitoring	Complex Problem Solving
Reading Comprehension	Critical Thinking
Speaking	Coordination
Coordination	Management of Financial Resources
Critical Thinking	Management of Personnel Resources
Management of Personnel Resources	Speaking
Social Perceptiveness	Systems Evaluation
Active Learning	Negotiation
Complex Problem Solving	Reading Comprehension

Engineers
General Management

Fig. A.2 Comparison of skill areas between disciplines including counts [1]

- Social Perceptiveness
- Speaking
- Systems Analysis
- Systems Evaluation
- Time Management
- Writing

Figure A.2 shows a comparison of the base engineering disciplines, how often skill areas were identified across the engineering disciplines and skill areas required by management and executive positions. This figure is similar to Fig. 6 in the text, but also shows the count that the various skill areas were listed for the five engineering disciplines analyzed. Areas highlighted in orange represent engineering skill areas. Areas highlighted in green represent management skills that aren't typically possessed by engineers from a functional engineering perspective. This provides some quantitative data to show engineers areas that they may need additional knowledge, skills and training.

Table A.7 Work styles of engineering disciplines [1]

Mechanical engineers	Electrical engineers	Chemical engineers	Civil engineers	Industrial engineers
Analytical thinking	Attention to detail	Analytical thinking	Integrity	Attention to detail
Attention to detail	Persistence	Integrity	Dependability	Dependability
Innovation	Analytical thinking	Attention to detail	Analytical thinking	Integrity
Dependability	Initiative	Dependability	Attention to detail	adaptability/ flexibility
Cooperation	Cooperation	Achievement/ Effort	Initiative	Analytical thinking

Table A.8 Work styles of management and leadership [1]

Architectural and engineering managers	General and operations managers	Chief executives
Analytical thinking	Dependability	Leadership
Integrity	Integrity	Initiative
Dependability	Attention to detail	Persistence
Attention to detail	Leadership	Analytical thinking
Stress tolerance	Initiative	Achievement/Effort

Work Styles

As defined by O*NET, "work styles are personal characteristics that can affect how well someone performs a job" [1]. Work styles represent how you approach your work and the accomplishment of your tasks. Table A.7 shows work styles that engineers possess.

From a management perspective, Table A.8 shows work styles that an engineering manager, general manager and executive should possess.

The following list shows each of these work styles. See Appendix "Definition of O*NET Knowledge Areas, Skill Areas" for a full definition of the work styles.

- Achievement/Effort
- Adaptability/Flexibility
- Analytical Thinking
- Attention to Detail
- Cooperation
- Dependability
- Initiative
- Innovation

Engineers	Number of Engineering Disciplines
Analytical Thinking	5
Attention to Detail	5
Dependability	4
Integrity	3
Cooperation	2
Initiative	2
Achievement/Effor	1
Adaptability/Flexibility	1
Innovation	1
Persistence	1

Architectural and Engineering Managers
Analytical Thinking
Integrity
Dependability
Attention to Detail
Stress Tolerance

Engineers
General Management

General and Operations Managers	Chief Executives
Dependability	Leadership
Integrity	Initiative
Attention to Detail	Persistence
Leadership	Analytical Thinking
Initiative	Achievement/Effort

Fig. A.3 Comparison of work styles between disciplines including counts [1]

- Integrity
- Leadership
- Persistence
- Stress Tolerance

Figure A.3 shows a comparison of the base engineering disciplines, how often work-style areas were identified across the engineering disciplines and work styles required by management and executive positions. This figure is similar to Fig. 7 in the text, but also shows the count that the various work style areas were listed for the five engineering disciplines analyzed. Areas highlighted in orange represent engineering work style areas. Areas highlighted in green represent management work styles that aren't typically possessed by engineers from a functional engineering perspective. This provides some quantitative data to show engineers areas that they may need additional knowledge, skills and training.

Definition of O*NET Knowledge Areas, Skill Areas and Work Styles

Definition of O*NET Knowledge Areas

- **Administration and Management**—Knowledge of business and management principles involved in strategic planning, resource allocation, human resources modeling, leadership technique, production methods, and coordination of people and resources.
- **Administrative**—Knowledge of administrative and office procedures and systems such as word processing, managing files and records, stenography and transcription, designing forms, and workplace terminology.
- **Building and Construction**—Knowledge of materials, methods, and the tools involved in the construction or repair of houses, buildings, or other structures such as highways and roads.
- **Chemistry**—Knowledge of the chemical composition, structure, and properties of substances and of the chemical processes and transformations that they undergo. This includes uses of chemicals and their interactions, danger signs, production techniques, and disposal methods.
- **Computers and Electronics**—Knowledge of circuit boards, processors, chips, electronic equipment, and computer hardware and software, including applications and programming.
- **Customer and Personal Service**—Knowledge of principles and processes for providing customer and personal services. This includes customer needs assessment, meeting quality standards for services, and evaluation of customer satisfaction.
- **Design**—Knowledge of design techniques, tools, and principles involved in production of precision technical plans, blueprints, drawings, and models.
- **Economics and Accounting**—Knowledge of economic and accounting principles and practices, the financial markets, banking, and the analysis and reporting of financial data.
- **Education and Training**—Knowledge of principles and methods for curriculum and training design, teaching and instruction for individuals and groups, and the measurement of training effects.
- **Education and Training**—Knowledge of principles and methods for curriculum and training design, teaching and instruction for individuals and groups, and the measurement of training effects.
- **Engineering and Technology**—Knowledge of the practical application of engineering science and technology. This includes applying principles, techniques, procedures, and equipment to the design and production of various goods and services.
- **English Language**—Knowledge of the structure and content of the English language including the meaning and spelling of words, rules of composition, and grammar.

- **Law and Government**—Knowledge of laws, legal codes, court procedures, precedents, government regulations, executive orders, agency rules, and the democratic political process.
- **Law and Government**—Knowledge of laws, legal codes, court procedures, precedents, government regulations, executive orders, agency rules, and the democratic political process.
- **Mathematics**—Knowledge of arithmetic, algebra, geometry, calculus, statistics, and their applications.
- **Mechanical**—Knowledge of machines and tools, including their designs, uses, repair, and maintenance.
- **Personnel and Human Resources**—Knowledge of principles and procedures for personnel recruitment, selection, training, compensation and benefits, labor relations and negotiation, and personnel information systems.
- **Physics**—Knowledge and prediction of physical principles, laws, their interrelationships, and applications to understanding fluid, material, and atmospheric dynamics, and mechanical, electrical, atomic and sub-atomic structures and processes.
- **Production and Processing**—Knowledge of raw materials, production processes, quality control, costs, and other techniques for maximizing the effective manufacture and distribution of goods.
- **Public Safety and Security**—Knowledge of relevant equipment, policies, procedures, and strategies to promote effective local, state, or national security operations for the protection of people, data, property, and institutions.
- **Sales and Marketing**—Knowledge of principles and methods for showing, promoting, and selling products or services. This includes marketing strategy and tactics, product demonstration, sales techniques, and sales control systems.

Definition of O*NET Skill Areas

- **Active Learning**—Understanding the implications of new information for both current and future problem-solving and decision-making.
- **Active Listening**—Giving full attention to what other people are saying, taking time to understand the points being made, asking questions as appropriate, and not interrupting at inappropriate times.
- **Complex Problem Solving**—Identifying complex problems and reviewing related information to develop and evaluate options and implement solutions.
- **Coordination**—Adjusting actions in relation to others' actions.
- **Critical Thinking**—Using logic and reasoning to identify the strengths and weaknesses of alternative solutions, conclusions, or approaches to problems.
- **Judgment and Decision Making**—Considering the relative costs and benefits of potential actions to choose the most appropriate one.

- **Management of Financial Resources**—Determining how money will be spent to get the work done, and accounting for these expenditures.
- **Management of Personnel Resources**—Motivating, developing, and directing people as they work, identifying the best people for the job.
- **Mathematics**—Using mathematics to solve problems.
- **Monitoring**—Monitoring/Assessing performance of yourself, other individuals, or organizations to make improvements or take corrective action.
- **Negotiation**—Bringing others together and trying to reconcile differences.
- **Operations Analysis**—Analyzing needs and product requirements to create a design.
- **Reading Comprehension**—Understanding written sentences and paragraphs in work-related documents.
- **Science**—Using scientific rules and methods to solve problems.
- **Social Perceptiveness**—Being aware of others' reactions and understanding why they react as they do.
- **Speaking**—Talking to others to convey information effectively.
- **Systems Analysis**—Determining how a system should work and how changes in conditions, operations, and the environment will affect outcomes.
- **Systems Evaluation**—Identifying measures or indicators of system performance and the actions needed to improve or correct performance, relative to the goals of the system.
- **Time Management**—Managing one's own time and the time of others.
- **Writing**—Communicating effectively in writing as appropriate for the needs of the audience.

Definition of O*NET Work Styles

- **Achievement/Effort**—Job requires establishing and maintaining personally challenging achievement goals and exerting effort toward mastering tasks.
- **Adaptability/Flexibility**—Job requires being open to change (positive or negative) and to considerable variety in the workplace.
- **Analytical Thinking**—Job requires analyzing information and using logic to address work-related issues and problems.
- **Attention to Detail**—Job requires being careful about detail and thorough in completing work tasks.
- **Cooperation**—Job requires being pleasant with others on the job and displaying a good-natured, cooperative attitude.
- **Dependability**—Job requires being reliable, responsible, and dependable, and fulfilling obligations.
- **Initiative**—Job requires a willingness to take on responsibilities and challenges.
- **Innovation**—Job requires creativity and alternative thinking to develop new ideas for and answers to work-related problems.

- **Integrity**—Job requires being honest and ethical.
- **Leadership**—Job requires a willingness to lead, take charge, and offer opinions and direction.
- **Persistence**—Job requires persistence in the face of obstacles.
- **Stress Tolerance**—Job requires accepting criticism and dealing calmly and effectively with high-stress situations.

Reference

1. O*NET Resource Center (onetcenter.org) This page includes information from the O*NET 28.1 Database by the U.S. Department of Labor, Employment and Training Administration (USDOL/ ETA). Used under the CC BY 4.0 license. O*NET® is a trademark of USDOL/ETA

MIX
Papier aus verantwortungsvollen Quellen
Paper from responsible sources
FSC® C105338

If you have any concerns about our products,
you can contact us on
ProductSafety@springernature.com

In case Publisher is established outside the EU,
the EU authorized representative is:
Springer Nature Customer Service Center GmbH
Europaplatz 3, 69115 Heidelberg, Germany

Printed by Libri Plureos GmbH
in Hamburg, Germany